Advanced Technologies for Science and Engineering

(Volume 3)

Intelligent Technologies for Research and Engineering

Edited by

S. Kannadhasan

*Department of Electronics and Communication Engineering
Study World College of Engineering
Coimbatore, Tamil Nadu-641105
India*

R. Nagarajan

*Department of Electrical and Electronics Engineering
Gnanamani College of Technology
Namakkal
Tamil Nadu, India*

Alagar Karthick
Department of Electrical and Electronics Engineering
K.P.R. Institute of Engineering and Technology
Coimbatore-641407, Tamil Nadu
India

K.K. Saravanan
Department of EEE
Anna University- Thirukkuvalai Campus
Thirukkuvalai
India

&

Kaushik Pal
Laboratório de Biopolímeros e Sensores, Instituto de
Macromoléculas
Universidade Federal do Rio de Janeiro
(LABIOS/IMA/UFRJ)
Rio de Janeiro, RJ- 21941-901
Brazil

Advanced Technologies for Science and Engineering

(Volume 3)

Intelligent Technologies for Research and Engineering

Editors: S. Kannadhasan, R. Nagarajan, Alagar Karthick, K.K. Saravanan and Kaushik Pal

ISSN (Online): 2859–3029

ISSN (Print): 2840–3029

ISBN (Online): 978-981-5196-26-9

ISBN (Print): 978-981-5196-27-6

ISBN (Paperback): 978-981-5196-28-3

need for a court order if at any point you breach any terms of this License Agreement. In no event will any delay or failure by Bentham Science Publishers in enforcing your compliance with this License Agreement constitute a waiver of any of its rights.

3. You acknowledge that you have read this License Agreement, and agree to be bound by its terms and conditions. To the extent that any other terms and conditions presented on any website of Bentham Science Publishers conflict with, or are inconsistent with, the terms and conditions set out in this License Agreement, you acknowledge that the terms and conditions set out in this License Agreement shall prevail.

Bentham Science Publishers Pte. Ltd.
80 Robinson Road #02-00
Singapore 068898
Singapore
Email: subscriptions@benthamscience.net

CONTENTS

PREFACE

The book on Intelligent Technologies for Research and Engineering covers new research findings from academics. The book contains research from active researchers who are involved in the cooperation between businesses and a variety of intelligent technologies, such as those that may be used in the production and distribution of industrial goods, factory automation, and other fields. The theory, design, development, testing, and evaluation of all intelligent technologies relevant to different areas of industry and its infrastructure are the main topics of this book. All computational intelligence techniques applicable to industry, intelligent data science techniques applicable to business and management, intelligent network systems applicable to industrial production, intelligent technologies applicable to smart agriculture, and intelligent information systems for agriculture are some of the topics covered. Significant advancements in intelligent systems have occurred as a result of the exponential growth of modern technologies. As a result, there is more potential for advancements and new uses.

A vital source of academic content on the creation, deployment, and integration of intelligent applications across several sectors is the journal, Developments and Trends in Intelligent Technologies and Smart Systems. This book is ideally suited for researchers, engineers, computer scientists, academics, students, and professionals interested in the most recent applications of intelligent technologies, highlighting a variety of cutting-edge topics like enterprise modelling, remote patient monitoring, and service-oriented architecture. Moreover, The latest advances in the field of solidification research and the problems posed by the community in the 21st century in terms of processing and analysis have been discussed.

On behalf of the editors, we would like to offer our appreciation to everyone who took part. First and foremost, the authors, whose excellent work is at the core of the book, and we gratefully congratulate all those involved and wish them great success. We would like to take this time to thank our family and friends for their support and encouragement while we worked on this book. First and foremost, we offer all credit and respect to our almighty Lord for his bountiful grace, which enabled me to finish this book successfully. We would like to express our gratitude to the writers for their contributions to this edited book. We would also like to thank Bentham Science Publishers and its whole team for facilitating the work and providing us the opportunity to be a part of this work.

The content of this book is summarized as follows:

In Chapter 1, it has been discussed that adding a new wireless access point, especially in a busy user environment does not always boost Wi-Fi performance. There are situations where, even with Access Points (AP) in every classroom, students still have to cope with slow download speeds. These common Wi-Fi complications are caused by co-channel interference. Wi-Fi communication is a bit like a conversation. A good communicator not only depends on the ability to speak but also on how well he or she listens. The conversational confusion is compounded when two speakers are using a similar tone. The same holds true for Wi-Fi transmission. Two or more neighboring APs operating on the same channel can increase interference and drag down performance. This paper proposes a smart antenna technology, when a smart antenna AP detects a neighboring AP signal, it will automatically change its pattern to reduce interference and show fast and reliable transmission. It is just like cupping our hands around our mouths or ears to let ourselves shout more loudly or listen more clearly. The normal methods for signal detection for WLAN nodes have many false positives. Therefore, this study proposes a BPNN model that uses a PFMDMM system for signal classification to identify the best signal for a WLAN node. The experimental results show that this Decision-Making Model using a Parameterized Fuzzy Measures Decision-Making Model Based on Preference Leveled Evaluation Functions for signal classification can better predict the best signal for a WLAN node. It was found that the estimated results and the signal detection accuracy are almost the same as in actual ground measurements. The test team simulated co-channel interference just like we would encounter in a school, office, hotel or airport where many APs operate on the same channel. The proposed smart antenna AP consistently delivered the best throughput outperforming other aps by an average of 75% superior coverage and unbeatable performance.

In Chapter 2, the fundamental point of this paper is to give staggered validation in biometric frameworks. Multimodal validation gives more degree of confirmation than unimodal biometrics, which utilizes only one biometric information, for example, unique finger impression or face or palm print. In this paper, we are utilizing a unique finger impression of an individual as a watermark which is installed in the chosen surface areas of the face picture of that individual which is caught utilizing a camera. A strategy called Discrete Wavelet Transform (DWT) is utilized for this reason. There is a serious level of visual relationship among unique and watermarked face pictures. The exhibition of the proposed watermarking strategy has been assessed and contrasted with procedures like Peak Signal-with Noise Ratio (PSNR) and Mean Squared Error (MSE).

In Chapter 3, Today, underwater communication has become a hot issue in research on both undersea and deep-sea navigation, as well as in autonomous underwater vehicle management, and acoustic communication has been accounted for due to its flexibility and lower degree of attenuation. However, owing to influencing elements such as channel time changing circumstances, bandwidth measurements, delay longer propagations, and the greatest degree of Doppler spread, pressure conditions, and salinity level, establishing acoustic communication in real-time is much more difficult. With a new monitoring era of global physical entities and based on the efficient energy-efficient awareness and depth, a new agent-based multipath routing protocol has been proposed in this work including underwater sensor nodes and underwater gateways with an autonomous underwater vehicle (AUV). The clustering head in the impacted region of sensor nodes will gather and aggregate data using mobile agent-initiated routing algorithms for identifying numerous pathways, as well as parameters including hope counting, delay propagation, nodal energy, and channel quality. In this paper, an agent-based dynamic AUV traversal method is developed to increase the network's dependability and connection while reorienting the AUV's movement direction

In Chapter 4, the number of people using face masks has increased on public transportation, retail outlets, and the workplace. All municipal entrances, workplaces, malls, schools, and hospital gates must have temperature and mask checks in order for people to enter such places. The paper's goal is to find someone who is not wearing a face mask in order to control COVID-19. Conv Net may be used to recognize and classify images. The model depends on Conv Net to assess whether or not someone is wearing a mask. It is possible to identify an image's face by utilizing a face identification algorithm. These faces are then processed using Conv Net face mask detection. If the model is able to extract patterns and characteristics from photographs, it will be categorized as either "Mask" or "No Mask". With an accuracy rate of 99.85 percent, Mobile Net V2 is the most accurate when it comes to training data. MobilenetV2 correctly identifies the mask in "Mask" or "No Mask" video transmissions.

In Chapter 5, PH plays an important role in determining product quality in industries like various chemical, petrochemical, and petroleum refineries, fertilizer, pharmaceutical, and food industries, effluent treatment, and many other organic and inorganic plants. For instance, in any industrial wastewater treatment plant, the PH is monitored and controlled by manipulating the acid or base stream, which is a strong acid or strong base. Modern treatment plant involves physical and chemical precipitation/flocculation along with biological treatment in-aerators/trickle filters, membranes, *etc*, where the control of PH is the key factor for efficient treatment. In chemistry, PH is a measure of the acidity or basicity of an aqueous solution. Pure water is said to be neutral, with apH close to 7.0 at 25 degree Celsius. Solutions with a PH less than 7 are said to be acidic and solutions with a PH greater than 7 are basic or alkaline. PH measurements are important in medicine, biology, chemistry, agriculture, forestry, etc. By PH control, we mean to maintain the PH value during continuous operation at a specific desired value by manipulating the alkaline flow rate. Usually in most industrial applications, the desired value is chosen to be around 7. This is the safest value for portable water, utility water used in industry, or waste-disposed water.

In Chapter 6, **Aedes albopictus** is considered the primary threatening vector for affecting public health. The process of identifying specific transcripts in enhancing the growth factor in **Aedes albopictus** is the initiation towards the development of a therapeutic marker. It implicates the identification of a particular antagonist. The approach was on the reference-based analysis of the whole transcriptome to reveal the differentially expressed pattern of transcripts. Further research requires the mathematical modeling of gene regulation and differential expression.

In Chapter 7, the objective of the study was to identify a potential inhibitor for a bifunctional protein in Microcystisaeruginosa. The in-silico modeling of the Protein using the "TBM" module of "Galaxy Seok Lab" extended the execution of virtual screening using MTi open screen. Finally, the protein-ligand interaction was studied using LIGPLOT software for "Bifunctional Protein" in "Microcystis aeruginosa." The virtual screening revealed 7176 compounds from the drug library, and the "best fit" screening resulted in 1500 compounds. Among the 1500 compounds, the molecule MK-3207 showed a better affinity towards the bifunctional Protein with -11.3Kcal/mol binding energy.

In Chapter 8, the focal point of this study is to recreate and plan a hybrid system consisting of solar photo-voltaic, battery, and diesel generator and; to determine its optimized configuration into an off-grid hybrid structure to meet the electricity demand of an institutional area situated in Jaipur, Rajasthan, India. Various configurations have different specifications that are obtained to meet the load demand based on input parameters which are obtained from the pilot survey and main survey as well at a particular location. Various costing parameters such as per unit of cost and net present cost are estimated with the condition of meeting the maximum load demand. The HOMER (Hybrid Optimization Model for Electric Renewable) software is used for different simulation processes and finally, it has been found that the solar PV-battery-diesel generator hybrid system is an economical system to meet the electricity demand in which the cost of energy is obtained as 13.83 and Net Present Cost is 9.78 M with initial capital and operating costs of 4.20 M and 646,319 per year, respectively. The diesel fuel cost is obtained as 5,09,288 per year. Meanwhile, the electricity production and consumption are also estimated to be 1,09,040 kWh/year and 81,939 kWh/year with an unmet load of 1.77% only, respectively.

In Chapter 9, one of the most important issues in Wireless Multimedia Sensor Networks is the energy efficiency of object detection and image transmission. In-node object detection and tracking algorithms have been proposed in recent WMSN approaches. However, with a little effort, the WMSN will able to detect the presence and absence of objects in images. For the WMSN, a new approach for the above technique is suggested in this research. Instead of sending a whole image, this technique sends image parts. It ensures energy saving inside the node and minimum picture content which is transferred to the sink node. On the basis of in-node reconstructed and energy consumption picture PSNR, it is suggested that the technique is evaluated using (PSNR). In comparison to existing state-of-the-art methodologies, simulation results demonstrate that the suggested methodology saves 95 percent of node energy with a received picture PSNR of 46 dB.

In Chapter 10, the purpose of this study is to discover anomalies and malicious traffic in the Internet of Things (IoT) network, which is critical for IoT security, as well as to keep a watch on and stop the undesired traffic flows in the IoT network. For this objective, a number of researchers have developed several machine learning (ML) approach models to limit fraudulent traffic flows in the Internet of Things network. On the other side, due to poor feature selection, some machine learning algorithms are prone to misclassifying mostly damaging traffic flows. Nonetheless, further study is needed into the vital problem of how to choose helpful attributes for accurate malicious traffic identification in the Internet of Things network. As a solution to the problem, an Artificial Neural Network (ANN) model is proposed. The Area under Curve (AUC) metric is used to employ the cross-entropy approach to effectively filter features using the confusion matrix and identify effective features for the chosen Machine Learning algorithm.

In Chapter 11, a mixed signal quadrature demodulator was suggested in this study. In 90 nm CMOS technology, to get the desired frequency range, a quadrature VCO is employed. The fast speed is achieved with a three-bit ADC. Unused ADC construction components have been removed to conserve energy and space. Outputs obtained are used to meet the power needed in the mixed signal demodulator designed for multi-gigabit applications. QVCO, baseband AGC, frequency synthesizers, and IQ mixers are all part of the demodulator. This displays the highest level of integration while using the least amount of electricity. To sample the symbols at optimal SNR, the baseband modem included a mixed signal timing recovery loop based on the Gardner timing error detector.

In Chapter 12, this paper focuses on wire length reduction throughout the 3D floor layout stage. The 3D cell layout stage is part of the floor planning process. Previously, it was expected that an entire module would be placed on a single device layer. They do not consider how a module's cells may be dispersed across many device levels to reduce cable length. Each of the device layers is assigned to one of the cells that make up a module (a 2D module is converted into 3D module). To place cells in three dimensions, several constraints are used. The placement-aware constraints are a set of constraints that determine whether a 2D module may be turned into a 3D module. The vertical alignment of identical sub-modules owing to the same planar placement requirement is referred to as vertical constraint. The size of the solution will be reduced as a result of this. A 3D floor design module packing method is proposed by the author. Calculating the wire length and taking into consideration the feasibility requirement for a smaller solution area, 3D cells are arranged in an initial set of floor layouts. After finding the best floor design, the modules are packed using a packing algorithm, and the technique is finished. A placement-aware 3D floor design method is the name of the approach, which is developed in C++ and operates on Fedora Linux.

In Chapter 13, the design and fabrication of biomimetic underwater robotic fish are covered. A robot fish is a type of bionic robot that looks and moves like a real fish. Two motors, an Arduino microcontroller, Bluetooth, and a pump are required to complete the underwater robotic fish project. Motors are employed for quick forward and rotating motion, and the pump assembly aids in deep-water diving. In addition, sensors assist the robot in making intelligent judgments such as obstacle detection, direction shift, and so forth. Additionally, essential information such as live streaming, pressure, and temperature are provided. The innovative performance of the robot helps achieve the real motion of the fish, making the robot competent for the aquatic-based design that helps to reduce the complex structure without applications such as underwater exploration, oceanic supervision, pollution level detection, and military detection. This project is also beneficial.

In Chapter 14, it has been discussed that agriculture is one of the backbones of Indian economy. India is primarily an agricultural country. It plays an important role in the development of our nation. This project proposes an automatic irrigation; it maintains the moisture content present in the soil by an automatic irrigation system. This setup uses a capacitive soil moisture sensor v1.2 that measures the exact amount of soil moisture. It monitors soil properties such as temperature, humidity, soil moisture, and motor status. These parameters are measured using a soil moisture sensor, a DHT11 sensor, which is controlled by a NodeMCU that acts both as a microprocessor and as a server. It is possible to remotely control many farm operations from any part of the world through IoT.

S. Kannadhasan
Department of Electronics and Communication Engineering
Study World College of Engineering
Coimbatore, Tamil Nadu-641105
India

R. Nagarajan
Department of Electrical and Electronics Engineering
Gnanamani College of Technology
Namakkal
Tamil Nadu, India

Alagar Karthick
Department of Electrical and Electronics Engineering
K.P.R. Institute of Engineering and Technology
Coimbatore-641407, Tamil Nadu
India

K.K. Saravanan
Department of EEE
Anna University- Thirukkuvalai Campus
Thirukkuvalai
India

&

Kaushik Pal
Laboratório de Biopolímeros e Sensores, Instituto de Macromoléculas
Universidade Federal do Rio de Janeiro (LABIOS/IMA/UFRJ)
Rio de Janeiro, RJ- 21941-901
Brazil

List of Contributors

A. Celciya Effrin — Department of Computer Science and Engineering, Francis Xavier Engineering College, Tirunelveli, India

Devendra Kumar Doda — Department of Electrical Engineering, Vivekananda Global University, Jaipur, Rajasthan-303012, India

Firos A. — Department of Computer Science and Engineering, Rajiv Gandhi University, Rono Hills, Doimukh-791112, India

Harishchander Anandaram — Centre for Excellence in Computational Engineering and Networking, Amrita Vishwa Vidyapeetham, Coimbatore, Tamil Nadu, India

J. Zahariya Gabrie — Department of Electronics and Communication Engineering, Francis Xavier Engineering College, Tirunelveli, India

Jacob Abraham — Department of Electronics, B P C College, Piravom P.O 686664, Kerala, India

K. G. Parthiban — Department of Electronics and Communication Engineering, Dhaanish Ahmed Institute of Technology, Coimbatore, India

Mallika Pandeeswari R. — Department of Electronics and Communication Engineering, Francis Xavier Engineering College, Tirunelveli, India

M. Philip Austin — Department of ECE, Francis Xavier Engineering College, Affiliated with Anna University, 103/G2, Bypass Road, Vannarpettai, Tirunelveli, Tamil Nadu 627003, India

M. Gunasekaran — Department of Computer Science, Government Arts College, Dharmapuri-636705, India

N.M. Nithish — Department of Electronics and Instrumentation Engineering, Kongu Engineering College, Perundurai, Erode, Tamil Nadu, India

Nandhakumar A. — Department of Electronics and Communication Engineering, Dhaanish Ahmed Institute of Technology, Coimbatore, India

Prashanth N. A. — Department of Electrical and Electronics, BMS Institute of Technology and Management, Bangalore-560064, India

Prasanth Venkatareddy — Department of Electrical and Electronics, Nitte Meenakshi Institute of Technology, Bangalore-560064, India

Praywin Samson E. — Department of Electronics and Communication Engineering, Francis Xavier Engineering College, Tirunelveli, India

Preethi Sharon — Department of Electronics and Communication Engineering, Francis Xavier Engineering College, Tirunelveli, India

Rajiga S.V. — Department of Computer Science, Government Arts College, Salem, Tamil Nadu 636007, India

Raja M. — Department of Electronics and Instrumentation Engineering, Kongu Engineering College, Perundurai, Erode, Tamil Nadu, India

R. Indhu Rani — Department of Computer Science and Engineering, Francis Xavier Engineering College, Tirunelveli, India

R. Tino Merlin — Department of Computer Science and Engineering, Francis Xavier Engineering College, Tirunelveli, India

Ravi R. Department of Electronics and Communication Engineering, Francis Xavier Engineering College, Tirunelveli, India

Rajeesh Kumar N.V. Computer Science and Engineering, Amrita College of Engineering and Technology, Coimbatore, India

R. Kabilan Department of ECE, Francis Xavier Engineering College, Affiliated with Anna University, 103/G2, Bypass Road, Vannarpettai, Tirunelveli, Tamil Nadu 627003, India

S. Kannadhasan Department of Electronics and Communication Engineering, Study World College of Engineering, Coimbatore, Tamil Nadu-641105, India

Srinivasan P. Department of Electronics and Communication Engineering, Amrita College of Engineering and Technology, Coimbatore, India

Saravana Shankar B. Department of Electronics and Instrumentation Engineering, Kongu Engineering College, Perundurai, Erode, Tamil Nadu, India

Sadhurwanth D. Department of Electronics and Instrumentation Engineering, Kongu Engineering College, Perundurai, Erode, Tamil Nadu, India

Samuel M. Department of Electronics and Communication Engineering, Francis Xavier Engineering College, Tirunelveli, India

Seema Khanum Department of Computer Science, Government Arts College, Salem, Tamil Nadu 636007, India

Shargunam S. Department of Electronics and Communication Engineering, Francis Xavier Engineering College, Tirunelveli, India

T. Jerlin Department of Computer Science and Engineering, Francis Xavier Engineering College, Tirunelveli, India

U. Maheshwari Department of Computer Science and Engineering, Francis Xavier Engineering College, Tirunelveli, India

V. Harini Priya Department of Computer Science and Engineering, Francis Xavier Engineering College, Tirunelveli, India

V. Brindha Department of Computer Science and Engineering, Francis Xavier Engineering College, Tirunelveli, India

Veluchamy S. Department of Electronics and Communication Engineering, Sri Venkadeswara College of Engineering and Technology, Coimbatore, India

CHAPTER 1

A Fuzzy Based High-Performance Decision-Making Model for Signal Detection in Smart Antenna Through Preference Leveled Evaluation Functions

Seema Khanum[1], M. Gunasekaran[2], Rajiga S.V.[1] and Firos A.[3,*]

[1] *Department of Computer Science, Government Arts College, Salem, Tamil Nadu 636007, India*

[2] *Department of Computer Science, Government Arts College, Dharmapuri-636705, India*

[3] *Department of Computer Science and Engineering, Rajiv Gandhi University, Rono Hills, Doimukh-791112, India*

Abstract: In a densely populated area with many users, adding a new wireless access point may not necessarily improve Wi-Fi performance. There are times when students must deal with poor download rates even with Access Points (AP) in every classroom. Cochannel interference is the root cause of several typical Wi-Fi issues. A discussion may be compared to Wi-Fi communication. The capacity to communicate and listen properly are both essential for effective communication. When two speakers are speaking in a similar tone, the conversational uncertainty is exacerbated. Wi-Fi broadcasts are the same way. The interference and drag performance might be worsened by two or more nearby APs using the same channel. This study suggests a smart antenna technology. When a smart antenna AP finds a nearby AP signal, it will automatically alter its pattern to minimise interference and provide quick and reliable transmission. The same principle applies when we cup our hands over our lips or ears to enable us to yell or listen more clearly. There are a lot of false positives in the typical approaches for WLAN node signal recognition. The optimal signal for a WLAN node is therefore identified using this study's proposed BPNN model, which uses the PFMDMM system for signal classification. This Decision-Making Model Using Parameterized Fuzzy Measures has been shown *via* experiments. A WLAN node's optimal signal may be more accurately predicted using a decision-making model based on preference-leveled evaluation functions. The precision of the signal identification and the anticipated findings were found to be almost identical to those obtained from real ground measurements. The test team mimicked cochannel interference, which would occur in a setting with plenty of APs, such as a workplace, hotel, or airport. The suggested smart antenna AP regularly outperformed other apps by an average of 75% greater coverage and unmatched performance.

* **Corresponding author Firos A.:** Department of Computer Science and Engineering, Rajiv Gandhi University, Rono Hills, Doimukh-791112, India; E-mail: firos.a@rgu.ac.in

S. Kannadhasan, R. Nagarajan, Alagar Karthick, K.K. Saravanan & Kaushik Pal (Eds.)

Keywords: BPNN, Decision making model, Deep learning model, Preference leveled evaluation functions, Received signal strength, Reconstruction, Signal detection, Smart antenna.

INTRODUCTION

Antenna Engineering

An antenna has a source for its input and output. Essentially, the input is an electrical transmission. Essentially, the output is an electromagnetic free-space signal. These inputs and outputs are of diverse kinds [6]. The signal's electrical character is present at the input. The electromagnetic signal's wave nature, which is nothing more than waves in empty space, is present at the output. An antenna is simply a device that transforms an electrical signal into an electromagnetic wave that travels across open space. It creates a wave out of an electrical pulse. The transducer operates on that principle. An electrical amount is changed into a nonelectrical quantity by the transducer. Take a speaker as an example [1 - 5].

An electrical signal is the speaker's input: for instance, our mobile device's speaker. The signal has an electrical composition. However, the voice may be heard in the output, making it a sound. It is an electrically inert quantity. Consequently, our speaker is a transducer. Inside the mobile ship, it transforms the electrical signal from the phone into a sound wave. The electrical quantity is transformed into nonelectrical values *via* the transducer. The technique also works in the opposite situation since the antenna doubles as a transducer. For instance, the input of a microphone is a sound wave, which is a nonlectrical quantity. The microphone then generates a voltage based on the sound wave, which is electrical by nature. Therefore, the ideas might be combined. A transducer is any device that transforms electrical values into nonelectrical quantities or vice versa. A transducer is one of several different types. A sinusoidal current of the voltage signal may be used to analytically analyze the nature of electrical signals [6 - 10].

Electromagnetic waves are emitted by the antenna. Assuming the input electrical signal is sinusoidal in form, the antenna will not be able to radiate the signal since the DC voltage will remain constant throughout time. An antenna can only transmit or produce this electromagnetic wave if the electrical signal it receives is sinusoidal in nature. A medium is not necessary for the propagation of an electromagnetic wave [11 - 15]. It is capable of travelling across empty space and moves at the speed of light in a vacuum.

Every time we discuss antennas, we get the impression that they may be enormous, hefty, and metallic in composition. In the past, antennas were often made of metal, were huge, and unwieldy. However, a lot has changed. Now, the idea of an antenna may be applied to tiny scales. There are also nanoantennas that are invisible to the human eye [16 - 20]. These categorizations are based on several circumstances. For instance, the first is a dish antenna, whereas the second is a satellite-related space antenna. Signals may reach billions of kilometers using space antennas, which are fueled by solar energy. In such instances, a massive or enormous antenna would be used as the receiving antenna. The dish size in astronomy antennas is huge. It is very strong, with a circumference of like thousands of feet. It can send messages to far-off galaxies and stars that are billions of kilometres away [30]. Additionally, it can pick up signals from such far-off locations. The strength of the antenna increases with its growth. The third kind of antenna is called a Wi-Fi antenna. It is linked to a wireless router.

Machine Learning in Antenna Design

A method called machine learning uses a large quantity of data to generate a result. A mathematical model is the result [2]. These labelled or well-defined data sets may be used to run a machine learning system. Machine learning is very sophisticated in that it can also develop its own pattern to recognise the incoming data and then produce a mathematical model that predicts the future. We have another subset of machine learning, which is termed deep learning. As a result, deep learning is slightly more intelligent than machine learning. In other ways, it's more intense because it resembles how the human brain works. There are many billions of neurons in the human brain. One neuron serves no use. But these billions of neurons' combinations and interactions with one another are what matter [21–25]. In order to create the output data sets, the deep learning technique acts on the input data sets in a manner that replicates the multi-layered structure of our brain's neurons [13].

Algorithms that are trained using training data sets are at the core of machine learning. The AI system receives the input data and creates a mathematical model based on training data sets [15]. In situations when we lack a closed form of an analytical model of any system, it is thus highly helpful. In conclusion, the machine learning algorithm requires input data to work, and we must also train it using training data sets. Then a mathematical model is produced as the result. If we supply enough data sets for training, the caliber of the training data sets will decide how accurate our mathematical model is in the output [31]. If we use high-quality data sets during training, the result will be a mathematical model that is very accurate.

The three main kinds of machine learning algorithms are reinforcement learning, unsupervised learning, and supervised learning [34]. The training data sets in a supervised learning system are labelled. When the input and output data sets are contained in the training data, they are extremely well-defined. We make the difference that whatever is the input would also be the output. The combination of labeled and precisely specified input and output data is the training data set. This algorithm has been trained using high-quality data sets.

So, if we offer enough training data sets, supervised learning can produce this good predictive mathematical model. The training data sets used by unsupervised learning techniques consist of unlabeled data. No information about the data sets is available to us. It produces a pattern and clusters the unlabeled data in order to categorise it. It is perceptive enough to recognise and separate the similarities between the existing unlabeled data sets. In order to produce a mathematical model based on the clustering effect, it develops its own pattern [26–30]. The data sets would be divided into a number of clusters with a number of shared traits. For an appropriate mathematical model in an unsupervised learning technique, we need a lot of data sets. We don't have any training data sets for the reinforcement learning method. To learn, the creature must interact with the environment *via* an agent. For instance, the agent must interact with the surroundings. It would interact with the environment, provide this reward—which may be favourable or unfavorable. A reward is a positive value, whereas a punishment is a negative value.

In a different illustration, after engaging with the surrounding environment, two different sensations are produced. The agent constantly learns it based on this experience, whether it is good or negative. It gains knowledge of what is what, what would happen if it learned certain things, *etc*. It carries out its acts in accordance with this learning experience.

It creates its own experiences, and learns from them based on the good and poor ones, the reward and the punishment. This is very potent, which is why Google developed AlphaGo [31 - 34], a program that has mastered playing chess and is capable of defeating any chess master in the world.

Deep Learning *Versus* Machine Learning

Although these two phrases are often used interchangeably in the literature when discussing science, there is little distinction between the two. Deep learning is a subset of machine learning, which is the superset. Actually, they both fall under the same category of domain. But there is a little distinction. The distinction is that with machine learning, we have algorithms and training data sets, and we may use various algorithmic techniques to produce mathematical models. Machine learning is the process of creating models from data. However, with deep learning, the input and output are both data since the neural network at its heart has been educated using learning principles. It is not an example. The optimum possible combination of input parameters and output would then be generated based on the training data sets and its learning experience. A deep neural network, or DNN, is a multilayered structure that is implied by the word "learning." The complexity increases with the number of layers.

The node in a Deep Neural Network (DNN) is comparable to a neuron in the human brain. There are millions of neurons in our brains, and when they communicate with one another, electrical impulses are produced. We have several hidden layers and countless billions of connections between neurons, which are shown as arrows. We have interactions from the input to the first hidden layer. Then, we have this weighted matrix, or w1. The value that is provided as the input to the first hidden layer is determined using this matrix. There are various connections between the first hidden layer and the second hidden layer. Another weighted matrix that we have is called w2. We have several weighted matrices for each level of hidden layers. The input value for the next hidden layer is specified by these matrices. We now get the final weighted matrix at the output. These intricate node connections and weighted matrices combine to form the output. The way the deep neural network works is quite similar to how human brain functions. Together, trillions of neurons form a vast network. Sometimes the output is not what we desire. As a result, there is a little margin of error, denoted as delta. on a reverse manner, this margin of error provides feedback to what is on the hidden layer. The backpropagation algorithm is the name given to this technique. This helps to cut down on mistake. The weight of the matrices is changed based on this inaccuracy. The backpropagation process reduces this error margin so that we output the values that are closest to the ideal. Overfitting is one issue that this DNN has. For instance, it is overfitting if the training data set is closely followed.

A dropout is the answer to this overfitting issue. With the dropout process, only a subset of the hidden layer's nodes are trained. We choose any random node from the hidden layer, train it, and then use it. By doing this, we actually reduce the quantity of hidden layer nodes being trained concurrently. We reduce the overfitting issue in DNN in this way.

Application of Machine Learning Algorithm in Smart Antenna

In areas with a large number of APs, we see poor WiFi performance. Even after installing a new wireless access point, in particular, the Wi-Fi performance may not always improve. Interference on the ccochannel is one of the causes of this. It may cause interference and result in poor performance if many nearby APs are using the same channel. This article proposes a smart antenna technology in which a smart antenna AP recognises a nearby AP signal and uses a BPNN model, or PFMDMM (Parameterized Fuzzy Measures Decision Making Model clustering approach), to choose the optimal signal for a WLAN node.

The remainder of this essay is organised as follows: The history of Smart Antenna, the distinction between a conventional array and a beamforming array, a comparison of switching and adaptive array antennas, and some fundamental concepts about the preference levelled evaluation function approach to generate fuzzy measurements are all covered in Section 2 [5]. Section 3 provides the suggested parameterized fuzzy measures decision making model for the optimal signal identification in smart antennas using the back propagation neural network (bpnn) method, which is based on preference levelled evaluation functions [14]. In order to test our proposed method, we obtained and preprocessed the cu/antenna dataset (v. 2009-05-08) from Community Resource for Archiving Wireless Data (CRAWDAD). The experiment is described in sSection 4, and Section 6 contains the study's main findings and conclusions.

THE BACKGROUND

The Smart Antenna

The smart antennas are also known as adaptive array antennas [24]. As the name indicates, the adaptive and array antenna elements are put together to form an antenna array. Adaptive means adjusting the pattern of the array. So, its name is adaptive array antenna. It is also recently called MIMO (Multiple Input Multiple Output Antenna Arrays), with smart signal processing algorithms used to identify the spatial signal signature which is the direction of the arrival signal [26]. The arrays are used in signal processing algorithms for the detection of the direction of arrival of the signal [3]. These are used to calculate beamforming vectors which are used to track and locate the antenna beam on the mobile target. Smart antennas are different from reconfigurable antennas. Most of the time, it happens that reconfigurable antennas are confused with smart antennas. But they are different from smart antennas. Both of them have similar capabilities. Single-element antennas are called reconfigurable antennas and not antenna arrays [5]. In regard to antenna arrays, they are smart antennas, but when it comes to simple antennas, we have reconfigurable antennas. The objective is to maximize the gain in the desired direction. Another objective is to minimize the gain in the direction of interference. In the direction of interference, we are going to nullify or minimize the radiation. However, in the direction of greatest benefit, the real objective is present [4] in that direction. We aim for maximum profit. Therefore, this is the intended user for our smart antenna when we indicate that we want the highest gain in the chosen direction. We thus seek to gain the most in this direction. As opposed to interference, where we desire a null in the case of the interfering users. A variety of components created with adaptively processed signals are used in the technical and objective functioning of smart antennas. The signals that undergo space-time filtering while being adaptively processed one after another [8, 29]. Adaptive array antennas and switching beam array antennas are the two main categories of smart antennas. The key part of mobile networks is the adaptive array antenna [29]. It enables the antenna to guide the beam in any desired direction while simultaneously cancelling the disruptive signal. In the sense of adapting, the beam is allowed to travel or directed in the desired direction, while the beam is suppressed or silenced in the direction of conflicting signals. The signals may be tracked and located using this adaptive antenna array. The adaptive algorithm is a component of smart antennas. It produces a beamforming signal as its output. When the adaptive algorithm receives the input signal, it will adjust such that it offers the highest gain in the desired direction and zero in the interfering direction. As a result, it can find and follow user signals, which means both users and interference. For each unique user, it may create a tailored radiation pattern. As a result, it may design or alter the radiation pattern for each user.

The Difference between Conventional Array and Beamforming Array

Each antenna has a unique radiation pattern. This is a whole antenna radiation pattern if we see them all. This is the situation with a typical array. However, if we claim that this is a result of the beamforming array, the radiation will rise or the gain will be greater in the intended direction. The target is located in the main beam. In such a situation, the primary beam and greater gain are visible. In contrast, the side lows in the opposite directions are visible and have very little increase. In terms of radiation, this is how the traditional array and the beamforming array vary.

A switch to beam array antenna is another kind of antenna. Turn the beam on and off as the name suggests. That indicates that the radiation beam is changing. Therefore, the switch beam array antenna contains a number of fixed beam patterns, and only the choice of which beam is to be accessible at a certain moment in time has to be made [33]. The switched beam system's goal is to improve gain in accordance with the user's position. We boost the gain of that specific beam based on the user's location. Let's illustrate it schematically here. Fig. (**1**) depicts the several beams that are present. However the node only anticipates one active beam. The antenna array is what causes these beams to form [21]. This beam is active at a certain time if a user is present. So there are many fixed beams, each with one pointed toward the required signal. In other words, the switch between the antenna, the beam array, or a single beam is directed towards the desired signal. One beam is being directed towards the intended signal.

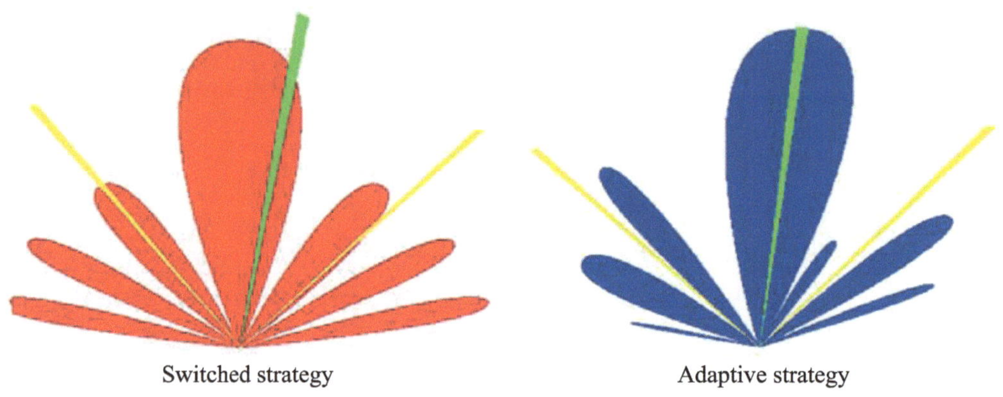

Fig. (1). Difference between switched beam and adaptive beam.

Comparison between Switched and the Adaptive Array Antenna

The maximum gain for the desired user, for interfering users, or for interference will be provided to a switched beam array antenna [1]. There are small loops, but in case of the adaptive scheme, the maximum gain is present for the desired user. However, there is also a null for users who are interfering. The adaptive plan looks like this. This is how switched antennas and adaptive antennas differ from one another. We can distinguish between the switch to beam and adaptive array based on the criteria [32]. The beam is simple to use and inexpensive to implement based on the integration switch. Trans receiver complexity occurs in the adaptive antenna array system, however, it is expensive and difficult. Based on range and coverage, adaptive arrays have less hardware redundancy. In contrast, the switch to beam array offers greater coverage than a traditional system and less coverage than an adaptive array [25]. Comparing adaptive arrays to switched beam systems, more coverage is available. In terms of interference rejection, there is a challenge in the switched beam in discriminating between the intended signal and the interferer and it doesn't respond to the moment of interference. Whereas in the adaptive array in the lobe, also the focusing is narrower and it is capable of nulling the interfering signals. It is capable of nullifying or nulling the applications. Mobile communications, cellular and wireless networks, satellite communications, wireless sensor networks, military applications, and electronic warfare are among the various applications. These are different applications of a smart antenna. Fig. (**1**) provides an illustration of how the switched beam and the adaptive beam differ from one another.

Preference Leveled Evaluation Functions Method to Construct Fuzzy Measures

This section discusses a method to construct parameterized fuzzy measures based on preference leveled evaluation functions. The practical meanings of some parameters can also be found in the numerical instance that follows.

This study considered 2.4 GHz (802.11b/g/n/ax) range WLAN. Spaced with 5mhz, fourteen channels are there in 2.4 GHz (802.11b/g/n/ax) range WLAN. 14 channels of 2.4 GHz range are given in Table **1**. A representation of overlapping channels within the 2.4 GHz band is given in Fig. (**2**).

Table 1. 14 channels of 2.4 GHz range.

Channel(x)	Frequency(F) (MHz)	Frequency Range
1	2412	2401–2423
2	2417	2406–2428
3	2422	2411–2433
4	2427	2416–2438
5	2432	2421–2443
6	2437	2426–2448
7	2442	2431–2453
8	2447	2436–2458
9	2452	2441–2463
10	2457	2446–2468
11	2462	2451–2473
12	2467	2456–2478
13	2472	2461–2483
14	2484	2473–2495

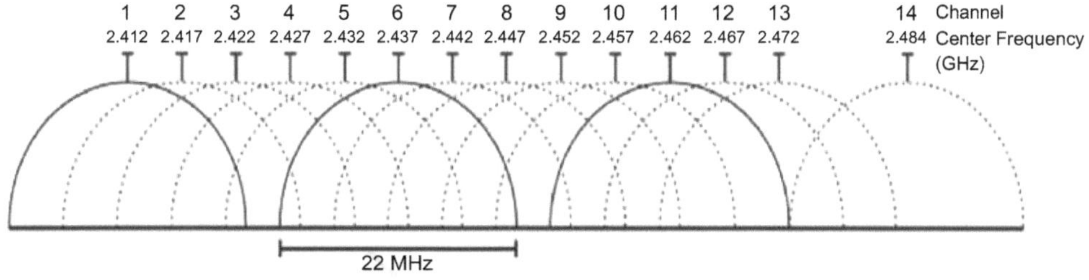

Fig. (2). 2.4 GHz band overlapping channels.

Construction Method Using Preference Leveled Evaluation Functions

With a normalized weights function $b = B^{<x>}$ having the complementary preference $\hat{\lambda}_b \in [0,1]^x$, and let $(F, 2^F, t_x)$ be a given fuzzy Frequency measures pace where $F = \{x(1), ..., x(n)\}$. The x values, that is the channel values in 2.4GHz range are given in Table 1. Here B is the band width and $B^{<x>}$ is the x-ary aggregate function. t_x is thespace of all fuzzy measures on X [16, 19].

Let us identify a parameterized fuzzy measure $t_G^{\varphi_1,......,\varphi_s}$ for parameterized fuzzy measure space $(G, 2^G, t_G^{\varphi_1,......,\varphi_s})$ where $G = \{1, ..., x\}$ [7].

We denote by $([0, 1]^n)^G$ the space of all mappings,

$$\varphi : G \to ([0, 1]^n) \tag{1}$$

We denote parameterized fuzzy measure $t_G^{(b,\varphi)} : 2^G \to [0, 1]$ such that for any $C \in 2^G (C \neq \varphi)$, we have

$$t_G^{(b,\varphi)}(C) = t_F(\{f \in F \mid (\vee_{g \in C} \varphi(g))(f) \geq \hat{\lambda}_b(|C|)\}) \tag{2}$$

and still define $t_G^{(b,\varphi)}(\varphi) = 0$ for all b and φ. This is adjustability conferring to an expected preference b.

Different complementary preference $\hat{\lambda}_b$ aid as diverse thresholds to be surpassed by union assessments $\vee_{g \in C} \varphi(g)$.

Moreover, when $C = G$, it follows $\{f \in F \mid (\vee_{g \in C} \varphi(g))(f) \geq \hat{\lambda}_b(|C|)\} = F,$

$$\text{then } t_G^{(b,\varphi)}(G)) = 1, \text{ always.} \tag{3}$$

Usually, the greater conservativeness, the lesser the parameterized fuzzy measure $t_G^{(b,\theta)}$, but it is not always the case. Usually, we will get the expected results, if band b', the two preferences satisfy some special relation.

$(B^{<x>}, \prec)$ gives a whole lattice with ordering $\prec$ demarcated such that for any two $b, b' \in B^{<x>}, b \prec b' \Leftrightarrow \lambda_b \prec \lambda_{b'}$ [16].

For any two $t_G^{(b,\varphi)}: 2^G \to [0,\ 1] \wedge_{b'}(j)$ for all $j \in \{1, ..., x\}$. So, the relation about conservativeness [17], *cn*.

$$cn(b) = \sum_{j=1}^{x} b(j) \cdot \frac{j-1}{n-1} = \frac{1}{x-1}\sum_{u=1}^{x}\sum_{m=s}^{x} b(m) = \frac{1}{x-1}\sum_{u=1}^{x}\left[1 - \sum_{m=1}^{u-1} b(m)\right]$$

$$= \frac{1}{x-1}\sum_{u=1}^{x}\left[1 - b(u) - \sum_{m=1}^{u} b(m)\right] \tag{4}$$

$$= \frac{x}{x-1} - \frac{1}{x-1} - \frac{1}{x-1}\sum_{u=1}^{x}\sum_{m=1}^{u} b(m)$$

$$= 1 - \frac{1}{x-1}\sum_{u=1}^{x}\sum_{m=1}^{u} b(m) \geq 1 - \frac{1}{x-1}\sum_{u=1}^{x}\sum_{m=1}^{u} b'(m) = cn(b')$$

Also,

for anytwo C, C' $\in 2^G$ with C $\subset$ C', $t_G^{(b,\varphi)}(C)) \leq t_G^{(b,\varphi)}(C'))$; $\tag{5}$

for any two b, b' $\in B^{<x>}$ with b $\prec$ b', $t_G^{(b,\varphi)}(C) \leq t_G^{(b',\varphi)}(C);$ $\tag{6}$

for any two φ, $\varphi' \in 2^F)^G$ with $\Theta < \Theta'$ (here "$<$" is φ(j)$<\varphi'$(j) for all $j \in$ G),

$$t_G^{(b,\varphi)}(C) \ \leq \ t_G^{(b,\varphi')}(C). \tag{7}$$

Readers may refer to the paper [10] for proof of the equations (1) to (7).

PROPOSED MODEL

Fig. (3) shows the Back Propagation Neural Network (BPNN) technique used in the proposed Parameterized Fuzzy Measures Decision Making Model based on PreferenceLeveled Evaluation Functions for Best Signal Detection in Smart Antenna. The model is given input channels from 2.4 GHz range signals to determine which channel is optimal for the node. The range-based received signal strength indicator (RSSI) approach is used to determine the signal range. Then, to find the best signals for the node, the Parameterized Fuzzy Measures Decision Making Model (PFMDMM) clustering approach is used. This method arranges the signals in the right way so that they may be fed into an artificial neural network for classification. During the iterative learning method with the aid of the backpropagation algorithm, the signal weights were supplied by an arbitrary value according to the x values in Table **1** and were modified for optimum performance. To determine if the acquired system accurately classifies the signal with the best-preferred signal and other signal portions, the neural network is evaluated against a number of test samples of signals during the testing phase.

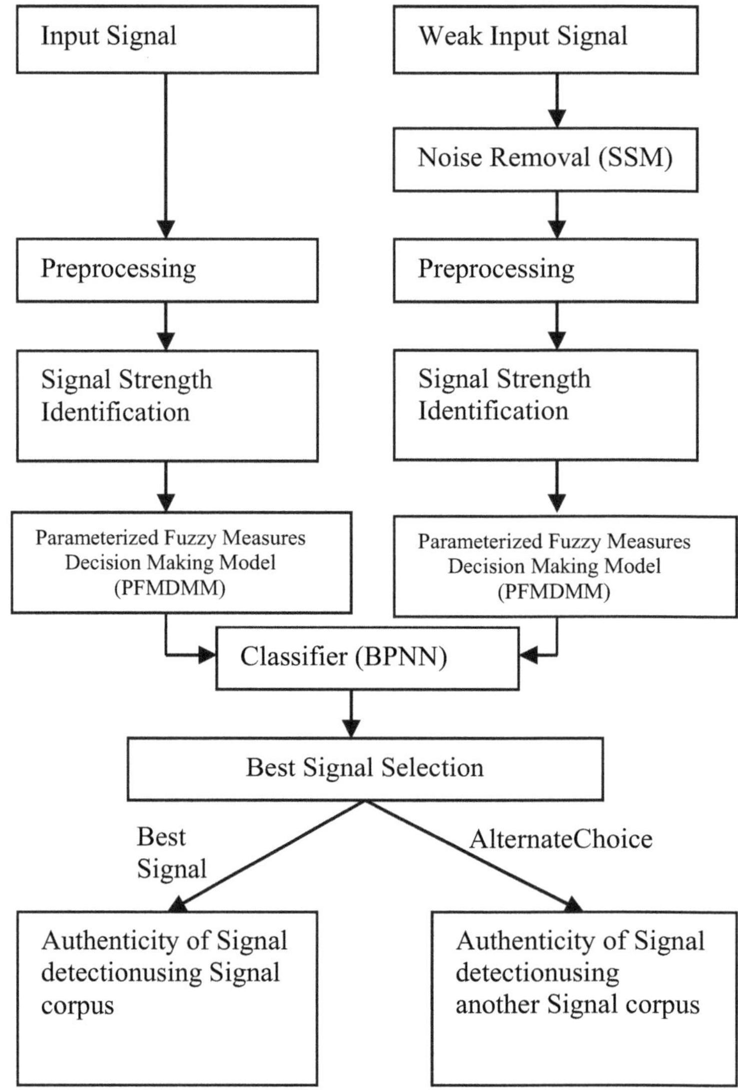

Fig. (3). Block diagram of proposed parameterized fuzzy measures decision making model based on preference leveled evaluation functions for best signal detection in smart antenna.

<table>
<tr><td>Algorithm 1: Decision Making Model Based on Preference Leveled Evaluation Functions for Best Signal Detection in Smart Antenna</td></tr>
<tr><td>

Input: Input channel of 2.4 GHz range signals.
Output: Categorizes the signal to the best-preferred signal and other signal parts.
Start
1. Input channels of 2.4 GHz range signals
2. Signal range detection with range-based received signal strength indicator (RSSI) method.
3. Parameterized Fuzzy Measures Decision Making Model clustering method (PFMDMM) to identify the best signals for the node; arranged in A proper format to feed into the Artificial Neural Network for classification.
4. *Training stage:* weights of the Feed Forward Neural Network were given by some arbitrary value as per Table -1 and were then tuned to optimal during the iterative learning procedure with the help of backpropagation algorithm.
5. *Testing stage:* The neural network is tested on a variety of test samples of signals, to ensure whether the acquired system correctly categorizes the signal into the best-preferred signal and other signal parts.
6. Categorizes the signal to the best-preferred signal and other signal parts.
Stop

</td></tr>
</table>

Signal Range Detection

Readers can refer to the steps for Signal range detection with range-based received signal strength indicator (RSSI) method in [28].

Parameterized Fuzzy Measures Decision-Making Model clustering method (PFMDMM) for Signal Detection

with $\varphi: G \rightarrow [0, 1]^4$, assume the normalized weights function $b \epsilon B{<}3{>}$ is assigned to be $w = (0.1, 0.4, 0.6)$.

We calculate $\lambda_b = (0.1, 0.3, 1)$ and $\hat{\lambda}_b = (0.9, 0.6, 0)$, with $cn(b) = cn\,(\lambda_b) = con(\hat{\lambda}_b) = 0.75$ showing a fairly added conservative measure in our preference. Then, using (2) we have for t_G :

$$t_G^{(b,\varphi)}(\{1\}) = t_F(\{f \epsilon F \,|(\varphi(1))(f) \geq \hat{\lambda}_b(1)\}) = t_F(\{4\}) = 0.2,$$

$$t_G^{(b,\varphi)}(\{2\}) = t_F(\{f \epsilon F \,|(\varphi(2))(f) \geq \hat{\lambda}_b(1)\}) = t_F(\{4\}) = 0.4,$$

$$t_G^{(b,\varphi)}(\{3\}) = t_F(\{f \epsilon F \,|(\varphi(3))(f) \geq \hat{\lambda}_b(1)\}) = t_F(\{3\}) = 0.4,$$

$$t_G^{(b,\varphi)}(\{1,2\}) = t_F(\{f \epsilon F \,|(\varphi(1) \vee \varphi(2))(f) \geq \hat{\lambda}_b(2)\}) = t_F(\{1,2,3,4\}) = 1,$$

$$t_G^{(b,\varphi)}(\{1,3\}) = t_F(\{f\epsilon F \mid (\varphi(1) \vee \varphi(3))(f) \geq \hat{\lambda}_b(2)\}) = t_F(\{1,3,4\}) = 0.9,$$

$$t_G^{(b,\varphi)}(\{2,3\}) = t_F(\{f\epsilon F \mid (\varphi(2) \vee \varphi(3))(f) \geq \hat{\lambda}_b(2)\}) = t_F(\{1,3,4\}) = 0.9,$$

$$t_G^{(b,\varphi)}(\{1,2,3\}) = t_F(\{f\epsilon F \mid (\varphi(1) \vee \varphi(2) \vee \varphi(3))(f) \geq \hat{\lambda}_b(3)\}) = t_F(\{1,2,3,4\}) = 1.$$

Deep Learning Model for Best Signal Selection

A deep neural network, often known as a Dnn, is essentially used in deep learning
or machine learning. After providing certain input parameters to the DNN with
the training data sets and learning rules, we get the output, which is a collection of
optimised parameters. If we wish to use this deep learning idea in the construction
of smart antennas, we may establish five parameters or variables at the input side:
inp=w, l, h, r, f = "patch width," "patch length," "thickness of the dielectric
substrate," "dielectric constant," and "operating frequency." We should anticipate
an optimised value of these five parameters if we feed this deep neural network's
input (inp) data. Be aware that the deep learning network will only function
successfully if we offer enough training data, since it will learn based on training
data that are of very high quality. In order to train the BPNN with the
anticipated/best permeates for every access point, the suggested model's
implementation employed the cu/antenna dataset (v. 2009-05-08) supplied by
community resources for archiving wireless data. In the proposed model, the
optimal signal for the access point is suggested by the BPNN (back propagation
neural network).

EXPERIMENTAL RESULTS

With the aid of the PFMDMM algorithm developed in MATLAB, the signal
vectors in the research are divided into the best-preferred signal and other signal
components. The model is given input channels from 2.4 GHz range signals to
determine which channel is optimal for the node. The range-based received signal
strength indicator (RSSI) approach is used to determine the signal range. Then, to
find the best signals for the node, the Parameterized Fuzzy Measures Decision
Making Model (PFMDMM) clustering approach is used. This method arranges
the signals in the right way so that they may be fed into an artificial neural
network for classification. During the iterative learning method with the aid of the
back propagation algorithm, the signal weights were supplied by an arbitrary
value according to the x values in Table **1** and were modified for optimum
performance. In order to determine if the acquired system accurately classifies the
signal to the best preferred signal and other signal portions, the neural network is
evaluated against a range of test samples of signals during the testing phase. For
quick and effective signal identification of an architecture, the PFMDMM is

employed. According to the research, the system is typically 94% effective in recognising signals.

Data Source

The Community Resource for Archiving Wireless Data (CRAWDAD) dataset, version 2009-05-08, is used in the research [11]. The signal strength data that was gathered to build a parametric model for 2.4 GHz directional 802.11 ad hoc antennas is included in this dataset. In order to improve these training data sets, we repeatedly tested this antenna for various parameters before producing a training data set from the simulation results. All dimensions were generated using commodity hardware, with the exception of the position patterns, by moving a large number of measurement packages between two antennas and recording the Received Signal Strength (RSS) at the conclusion. The three antenna configurations used are: an 8-degree horizontal beam-width HyperLink 24dBi parabolic dish; a 30-degree horizontal beam-width HyperLink 14dBi patch; and an 8-element uniform circular phased array from Fidelity ComtechPhocus 3000 with a main-lobe beam-width of roughly 52 degrees.

The data sets primarily include two pieces of information: normalised RSS (norm.rss) and packet received signal strength (RSS). The absolute RSS is known as norm.rss and is calculated by subtracting it from the trace's "reference maximum". The maximum reference value is set as the greatest mean value for any angle inside the trace. Table **2** provides a portion of the cu/antenna dataset's content (v. 2009-05-08).

Table 2. Example data of cu/antenna dataset (v. 2009-05-08).

Id	Rss	Batch	Position	Tag	Norm.rss	Norm.diff
1	48	parabolic-field2	0	default	-2.660532267	2.660532267
2	53	parabolic-field2	0	default	2.546716223	-2.546716223
3	51	parabolic-field2	0	default	0.463816827	-0.463816827
4	51	parabolic-field2	0	default	0.463816827	-0.463816827
5	50	parabolic-field2	0	default	-0.577632871	0.577632871
6	51	parabolic-field2	0	default	0.463816827	-0.463816827
7	50	parabolic-field2	0	default	-0.577632871	0.577632871
8	49	parabolic-field2	0	default	-1.619082569	1.619082569
9	50	parabolic-field2	0	default	-0.577632871	0.577632871
10	52	parabolic-field2	0	default	1.505266525	-1.505266525
11	51	parabolic-field2	0	default	0.463816827	-0.463816827

(Table 1) cont.....

Id	Rss	Batch	Position	Tag	Norm.rss	Norm.diff
-	-	-	-	-	-	-
8715891	1.237726	patty-field	225	default	-9.639062036	-82143414

Illustrative Example

The screenshot of the PFMDMM-based signal detestation simulation application is given in Fig. (**4**). The signal depicted in blue colour is the proffered signal suggested by the system for a node.

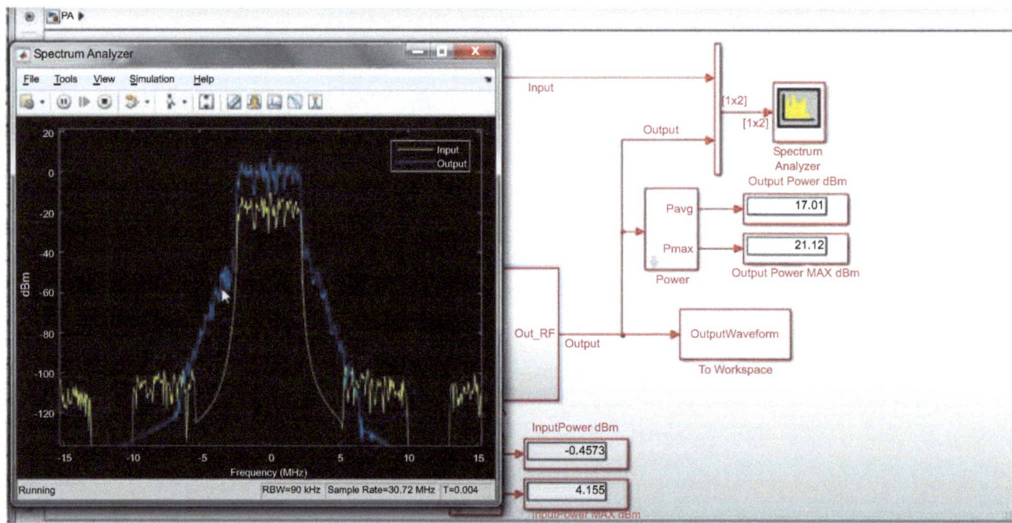

Fig. (4). PFMDMMbased signal detestation simulation application.

Experiments with Laboratory Data

In order to verify the performance of BPNN models, that is, using PFMDMM system for signal detection, experiments are realized on the laboratory data. For accurate signal detection metal, the determination accuracy, namely, Producer accuracy, User accuracy, and Overall accuracy were 91.50%, 95.32%, and 94.99%, respectively.

CONCLUSION

This work presents a unique BPNN model that employs the PFMDMM system for signal classification to determine the optimal signal for a node. The experiment findings show that the parameterized fuzzy measure decision-making model for Best Signal Detection in Smart Antenna based on preference leveled assessment function exhibits promising performance in terms of a best signal recommendation for a WLAN node.

To the best of our knowledge, this is the first research to employ the preference-leveled evaluation functions-based parameterized fuzzy measures decision-making model for signal categorization.

The primary contributions of this work are as follows:

• The suggested strategy is capable of providing satellite-based picture clustering decision-making within short time frames, which aids in more methodically addressing the geological problems.

• For the signal classification-based BPNN architecture for the feature extraction process, we present a novel application of parameterized fuzzy measures decision-making model based on preference levelled evaluation functions.
• In order to test the suggested technique, the experiment's cu/antenna dataset (v. 2009-05-08) has been taken into consideration. The signal strength data that was gathered to build a parametric model for (2.4 GHz) directional 802.11 ad hoc antennas is included in this dataset.

We specifically used the decision-making capabilities of the Parameterized Fuzzy Measures Decision Making Model Based on Preference Levelled Evaluation Functions to define a novel automated signal suggestion model.

REFERENCES

[1]   A. Asrokin, M.K.A. Rahim, and Aziz Abd, "Dual band microstrip antenna for wireless LAN application", *Proceedings of the 2005 asia pacific conference on applied electromagnetics,* Johor, Malaysia, pp. 10698-110701, 2015.

[2]   A. Mahmood, M. Bennamoun, S. An, F. Sohel, F. Boussaid, R. Hovey, G. Kendrick, and R.B. Fisher, "Deep Learning for Coral Classification", In: *Handbook of Neural Computation.* Academic Press, 2017, pp. 383-401.
[http://dx.doi.org/10.1016/B978-0-12-811318-9.00021-1]

[3]   B. Kadri, and F.T. Bendimered, "Fuzzy Genetic Algorithms for the synthesis of unequally spaced microstrip antennas arrays", *First European conference on antennas and propagation,* Nice, France, pp-1-6, 2006.
[http://dx.doi.org/10.1109/EUCAP.2006.4584907]

[4]   B. Kadri, and F.T. Bendimered, "Linear Antenna Synthesis with a Fuzzy Genetic Algorithm", *EUROCON 2007 - The International Conference on "Computer as a Tool",* Warsaw, Poland, 2007.
[http://dx.doi.org/10.1109/EURCON.2007.4400530]

[5]   B. Kadri, F.T. Bendimered, and E. Cambiaggio, "Modelisation of the feed network application to synthesis unequally spaced microstrip antennas arrays", *International Conference on Electromagnetics in Advanced Applications ICEAA 99,* Torino, ItalY, pp.371-374, 1999.

[6]   B. Aghoutane, S. Das, Mohammed EL Ghzaoui, B.T.P. Madhav, and Hanan El Faylali, "A novel dual band high gain 4-port millimeter wave MIMO antenna array for 28/37 GHz 5G applications", *AEU -*

Int. J. of Electronics and Commu., vol. 145, p. 154071, 2022.
[http://dx.doi.org/10.1016/j.aeue.2021.154071]

[7]   S. Bodjanova, and M. Kalina, "Approximate evaluations based on aggregation functions", *Fuzzy Sets Syst.,* vol. 220, pp. 34-52, 2013.
[http://dx.doi.org/10.1016/j.fss.2012.07.014]

[8]   Boyapati Bharathi Devi, and Jayendra Kumar, "Small frequency range discrete bandwidth tunable multiband MIMO antenna for radio/LTE/ISM-2.4 GHz band applications", *AEU - Int. J. Electron. Commun.,* vol. 144, p. 154060, 2022.
[http://dx.doi.org/10.1016/j.aeue.2021.154060]

[9]   C. Wei, T. Chen, and S. Lee, "K-NN based neuro-fuzzy system for time series prediction", *Proc. Int. Conf Software Engineering, Artificial Intelligence, Networking and Parallel/Distributed Computing,* Honolulu, pp. 569-574, 2013.
[http://dx.doi.org/10.1109/SNPD.2013.68]

[10]   C. Zhu, L. Jin, R. Mesiar, and R.R. Yager, "Using preference leveled evaluation functions to construct fuzzy measures in decision making and evaluation", *Int. J. Gen. Syst.,* vol. 49, no. 2, pp. 161-173, 2020.
[http://dx.doi.org/10.1080/03081079.2019.1668384]

[11]   W. Eric, "Anderson, Caleb Phillips, Crawdad dataset cu/antenna", Available at: https://crawdad.org/cu/antenna/20090508/rss

[12]   F. Li, and Y. Du, "From alphago to power system AI: What engineers can learn from solving the most complex board game", *IEEE Power Energy Mag.,* vol. 16, no. 2, pp. 76-84, 2018.
[http://dx.doi.org/10.1109/MPE.2017.2779554]

[13]   G. Angiulli, "Comments on "a hybrid method based on combining artificial neural network and fuzzy inference system for simultaneous computation of resonant frequencies of rectangular, circular, and triangular microstrip antennas", *IEEE Trans. Antenn. Propag.,* vol. 57, no. 1, pp. 296-296, 2009.
[http://dx.doi.org/10.1109/TAP.2008.2009786]

[14]   M. Grabisch, J.L. Marichal, R. Mesiar, and E. Pap, *Aggregation Functions.* Cambridge University Press: Cambridge, 2009.
[http://dx.doi.org/10.1017/CBO9781139644150]

[15]   Jie Zhang, and A.J. Morris, "Recurrent neuro-fuzzy networks for nonlinear process modeling", *IEEE Trans. Neural Netw.,* vol. 10, no. 2, pp. 313-326, 1999.
[http://dx.doi.org/10.1109/72.750562] [PMID: 18252529]

[16]   L. Jin, and R. Mesiar, "The metric space of ordered weighted average operators with distance based on accumulated entries", *Int. J. Intell. Syst.,* vol. 32, no. 7, pp. 665-675, 2017.
[http://dx.doi.org/10.1002/int.21869]

[17]   K. Wang, "A new fuzzy genetic algorithm based on population diversity", *International Symposium on Computational Intelligence in Robotics and Automation.* Banff, AB, Canada, pp. 108-112, 2001.

[18]   K. Das Ayinala, and P.K. Sahu, "Isolation enhanced compact dual-band quad-element MIMO antenna with simple parasitic decoupling elements", *AEU - Int. J. Electron. Commun.,* vol. 142, p. 154013, 2021.
[http://dx.doi.org/10.1016/j.aeue.2021.154013]

[19]   B. Llamazares, "Choosing owa operator weights in the field of social choice", *Inf. Sci.,* vol. 177, no. 21, pp. 4745-4756, 2007.
[http://dx.doi.org/10.1016/j.ins.2007.05.015]

[20]   M.A. Ashraf, and A.R. Khalid Jamil, "Evaluation of a single-input multiple-output antenna array for ultra-wide band applications", *AEU - Int. J. Electron. Commun.,* vol. 79, pp. 291-300, 2021.
[http://dx.doi.org/10.1016/j.aeue.2017.06.019]

[22]   P. Diniz, *Adaptive Filtering Algorithms and Practical Implementation.* 4th ed. Springer: New York,

2013, pp. 152-154.
[http://dx.doi.org/10.1007/978-1-4614-4106-9]

[23]   P.K. Guru Diderot, N. Vasudevan, and K.S. Sankaran, "An efficient fuzzy c-means clustering based image dissection algorithm for satellite images", *2019 International Conference on Communication and Signal Processing (ICCSP)*, Chennai, India, 2019.
[http://dx.doi.org/10.1109/ICCSP.2019.8698054]

[24]   N.G.R. Raimundo, C.A. Pitz, L.O.B. Eduardo, and R. Seara, "An ?0-norm-constrained adaptive algorithm for joint beamforming and antenna selection", *Digital Signal Processing.*, vol. 126, p. 103475, 2022.
[http://dx.doi.org/10.1016/j.dsp.2022.103475]

[25]   S. Chen, S. Liu, and L. Hanzo, "Adaptive bayesian space-time equalisation for multiple receive-antenna assisted single-input multiple-output systems", *Digital Signal Processing.*, vol. 18, no. 4, pp. 622-634, 2021.
[http://dx.doi.org/1016/j.dsp.2007.09.012]

[26]   S. Jaisiva, T. Rampradesh, and P. Vimala, "Epoxy based on low profile MIMO antenna for cellular systems", *Materials Today Proceedings.*, vol. 46 (part-9), pp. 4099-4101, 2021.
[http://dx.doi.org/10.1016/j.matpr.2021.02.6282]

[27]   S.R. Ostadzadeh, M. Soleimani, and M. Tayarani, "Prediction of induced current in externally excited dipole antenna using fuzzy inference", *2008 Second Asia International Conference on Modelling & Simulation (AMS)*, Kuala Lumpur, Malaysia, pp. 1039-1042, 2008.
[http://dx.doi.org/10.1109/AMS.2008.66]

[28]   J. Tang, and J. Han, "An improved received signal strength indicator positioning algorithm based on weighted centroid and adaptive threshold selection", *Alex Eng J.*, vol. 60, no. 4, pp. 3915-3920, 2021.
[http://dx.doi.org/10.1016/j.aej.2021.02.031]

[29]   T. Su, H. Hung, J. Cheng, C. Lu, and C. Hung, "Fuzzy theory application to the satellite antenna controller", *2014 International Conference on Information Science, Electronics and Electrical Engineering*, Sapporo, Japan, pp. 690-693, 2014.
[http://dx.doi.org/10.1109/InfoSEEE.2014.6947753]

[30]   A. Thiago, Bruza Alves, and Taufik Abrão, "Massive MIMO and NOMA bits-per-antenna efficiency under power allocation policies", *Phy. Commun.*, vol. 51, p. 101588, 2022.
[http://dx.doi.org/1016/j.phycom.2021.101588]

[31]   W. Orozco-Tupacyupanqui, H. Pérez-Meana, and M. Nakano-Miyatake, "A new method for searching the optimal step size of NLMS algorithm in intelligent antennas arrays based on fuzzy logic and artificial neural network", *IEEE Central America and Panama Convention (CONCAPAN XXXIV)*, Panama, Panama, 2014.
[http://dx.doi.org/10.1109/CONCAPAN.2014.7000421]

[32]   J. Wu, C. Liang, and Y. Huang, "An argument-dependent approach to determining OWA operator weights based on the rule of maximum entropy", *Int. J. Intell. Syst.*, vol. 22, no. 2, pp. 209-221, 2007.
[http://dx.doi.org/10.1002/int.20201]

[33]   Z. Xu, "On method for uncertain multiple attribute decision making problems with uncertain multiplicative preference information on alternatives", *Fuzzy Optim. Decis. Making.*, vol. 4, no. 2, pp. 131-139, 2005.
[http://dx.doi.org/10.1007/s10700-004-5869-2]

[34]   Y. Tawk, J. Costantine, S.E. Barbin, and C.G. Christodoulou, "A multi-band microstrip antenna design using cellular automata and fuzzy ARTMAP Neural Networks", *3rd European Conference on Antennas and Propagation* Berlin, Germany, pp. 3511-3514, 2009.

CHAPTER 2

Multimodal Biometric Authentication Using Watermarking

Shargunam S.[1,*], **Mallika Pandeeswari R.**[1], **Ravi R.**[1], **Preethi Sharon**[1], **Praywin Samson E.**[1] and **Samuel M.**[1]

[1] *Department of Electronics and Communication Engineering, Francis Xavier Engineering College, Tirunelveli, India*

Abstract: The primary goal of this study is to provide a robust validation for biometric systems. When compared to unimodal biometrics, which use just one biometric feature, such as a unique finger print, facial feature, or palm print, multimodal validation provides a higher level of confirmation. In this paper, we use an individual's distinctive fingerprint as a watermark that is installed in the desired locations of the facial image of that individual that is captured with the aid of using a camera. This is accomplished by using a technique known as the Discrete Wavelet Transform (DWT). Between unwatermarked and watermarked face images, there is a significant serious level of visual relationship. Peak Signal-to-Noise Ratio (PSNR) and Mean Squared Error (MSE) procedures have been compared to and evaluated by the proposed watermarking strategy.

Keywords: Biometrics, Discrete wavelet transform, Inverse discrete wavelet transform, Watermarking.

INTRODUCTION

Multimodal biometric frameworks have enabled ID recognition and security over the last several years. To increase the dependability of person recognition, a multimodal biometric framework mixes at least two biometric information acknowledgment results. Multimodal biometric frameworks use or have the ability to use a variety of physiological IDs or verifications. Several watermarking systems are available for safely encoding data in a picture. There are two different sorts of techniques that may be used: change space process and spatial area strategy. In this study, watermarking in connection to biometric identifiers is applied. Fingerprints and facial characteristics are valid biometric identifiers.

---

* **Corresponding author Shargunam S.:** Department of Electronics and Communication Engineering, Francis Xavier Engineering College, Tirunelveli, India; E-mail: shargunamguna@gmail.com

When compared to a photograph of a face, the fingerprints are smaller. As a result, we implant a unique finger impression in an individual's facial photograph to bend the extricated watermark.

LITERATURE SURVEY

Sub Band Splitting Technique for Embedding the Fingerprint

A logical biometric watermarking improves fingerprint security. Fingerprints are obtained by using a DWT technique to embed a face as a watermark in a unique finger impression. Fingerprints are exposed to DWT and afterwards, parted into sub-groups. A sign is deteriorated utilizing a 1-dimensional wavelet change to acquire the changed sign. A 2-dimensional wavelet change is acquired by applying a 1-dimensional change along the lines and sections. The 2-level disintegration comprises many subgroups like HH, HL, LH, and LL, and it contains the slanting, even, vertical, and coarse subtleties of the unique finger impression picture. When using Discrete Wavelet Transform to remove the inserted unique mark and text images, the first set of watermarked images is not required. Instead, we can install two watermarked images: a unique finger impression image and related text information of a person into the chosen surface areas of finger impression images. The absence of the watermark in the removed watermark would reveal that the integrity of the facial image has been jeopardised.

However, we will merely insert the unique mark picture into the face image in this instance. The face picture is broken into smaller groups after the DWT correction. We are unable to implement any improvements at this level since the LL2 already handles the finer details of the image. There will be adjustments in the first image if any progress is made in this area. When that happens, we may implant the finger imprint in the subgroups HL2, LH2, and HH2 since only minor details of the facial image are present in these subgroups.

In the unlikely event that any development is made in this area and none is discovered, we may look for progress in the face image. Accordingly, we can install the finger imprint image in these subgroups. Then, at that time, we may store any text-based repeated information in the subgroups HL1, HH1, and LH1. On separating the surface components of the face picture, we must first get the face image. Next, we must apply DWT on the face image, which will result in several subgroups.

The finger imprint image may then be added at that point. IDWT may then be used to get the watermarked facial image. The watermarked face image may then be stored in the data set at that moment. Currently, we will apply the DWT modification to the watermarked face image. The sub-band will then be produced, indicating that the surface elements of the watermarked face image may be made at that moment. The finger imprint image that will be injected into the facial image may be removed with the use of a watermark extractor approach. As a result, the unique mark image may be divided [1 - 5].

Maintaining the Integrity of the Specifications

A technique for identification and access control is biometric authentication, sometimes referred to as multimodal biometric authentication. This technique may also be used to identify members of organisations. Biometric identifiers are often divided into categories based on physiological and behavioural traits. A physical biometric is one that is based on an individual's physical characteristics. Examples include fingerprints, hand geometry, retinal scans, and DNA. Examples of behavioural biometrics include speech patterns, signatures, and keystrokes. A behavioural biometric is one that is based on an individual's behaviour. Since biometric identifiers are more trustworthy than tokens in confirming identification. The signatureA multimodal biometric system may combine fingerprint and facial recognition. A system that may use more than one physiological or behavioural trait is called a multimodal biometric. To provide the highest security, multimodal biometrics combines the aforementioned recognition technologies. If one of the technologies malfunctions for whatever reason, it may still accurately identify a person. Utilising many forms of identity, maintaining a high threshold of recognition, and allowing the system administrator to choose any degree of security are all advantages of multimodal biometrics. Up to three biometric identities may be used for a site with very high security. This significantly lowers the likelihood of accepting a forger. Multimodal biometrics are increasingly necessary since unimodal biometrics only use one kind of identification.

Examples include face, fingerprint, and voice recognition systems, which are susceptible to issues like noisy data, non-universality, and spoofing assaults. It results in a high rate of erroneous acceptance and false rejection. Compared to unimodal systems, multimodal systems are more dependable. Using multimodal biometrics has a number of advantages, including enhancing the entire system's accuracy and providing a backup method of enrolment and identification in the event that not enough information can be gleaned from a particular biometric sample. It can also detect efforts to trick biometric systems using non-live data sources, including phoney fingers [6 - 10].

Problem Description

Human physiology in several areas may be used for verification. Seven of them will be used to assess the reasonableness of the characteristic for validation while weighing the various factors to distinguish when selecting the specific biometric for a given application. Comprehensiveness means that any person using a framework should possess the characteristic. Being unique indicates that each person's characteristics must be unique so they may be distinguished from one another. The characteristic changes over time are referred to as lastingness. Quantifiability is related to how the attribute is estimated. The acquired information should also enable the subsequent preparation and extraction of the pertinent highlights. Execution is related to confirmation accuracy, check speed, and the capability of the used framework. Adequacy refers to how well the general public accepts the system to the point where their attributes are seen and evaluated. If all other factors are equal, no other unimodal biometrics will be able to satisfy all requirements. Due to the imperfect obtaining conditions in unimodal biometrics, the captured biometric information may be misinterpreted and the chosen client of the framework may be wrongfully rejected. Unimodal biometrics has a tendency to identify similarities in large populations. A facial acknowledgment camera may not be ready to recognize the two indistinguishable twins. Because of the blurred prints or immature unique finger impressions, the edges of older individuals and little youngsters' figure impressions might experience issues in taking a crack at a finger impression framework. Unimodal biometric frameworks are powerless against parody assaults. Thus, if by some stroke of good luck one biometric information, for example, a unique finger imprint or face, is applied, it may induce attack without any trouble. To access to the information that is input to the PC or on the circle, the data is incapable of duplication from some other frameworks by using programming [11 - 15].

PROPOSED MULTIMODAL BIOMETRIC SYSTEM

The suggested multimodal biometric system is discussed in this section. The proposed algorithm and watermarking method are described in detail.

Watermarking

A common technique for obscuring messages in images, sounds, or video files that can't be deleted or replaced is watermarking. A watermark is a recognisable image or example on the paper that is used to indicate validity. There are many uses for watermarks, including making it difficult to falsify government documents and postage stamps. To safeguard and validate biometric data and improve the accuracy of recognition, many watermarking techniques have been used. In essence, watermarking is used to ensure authenticity. The two different watermarking techniques used for securely embedding information in a picture are the change area procedure and the spatial space method. A kind of marking that is often used to denote copyright ownership is a computerised watermark. The signal in this case alludes to the face image. Multiple watermarks may be displayed on a sign at once; a computerised watermark has no effect on the size of the transporter signal. The transporter signal can have an advanced watermark applied to it to stamp media documents with copyright information. If everything were equal, a delicate watermark would be used to ensure accuracy. Steganographic techniques are used in advanced watermarking and steganography to covertly add information to loud signals. At each point of conveyance, a watermark is thus added to a computerised signal. Although in some instances the term "computerised watermark" refers to the distinction between the watermarked signal and the cover signal, advanced watermark refers to the data that will be added to a sign. The signal where the watermark will be inserted is called the host signal. The three stages of watermarking are typically described as implanting, attacking, and positioning. The watermarked computer signal is then transmitted to a different recipient. Assuming that the individual makes any change, it is called an assault. At the point when the adjustment by any individual may not be pernicious, then, at that point the term assault might come from a copyright insurance application. There are numerous changes that happen, for instance, lossy compression of the information, and trimming a picture. To distinguish the watermark from the attacked sign, identification is a calculation that is applied to the target sign. Assuming the sign was not altered during transmission, the watermark is accessible and it tends to be extricated. Even if the modifications are very strong in digital watermarking applications, the extraction algorithm should produce the watermark correctly. If the signal is altered in fragile digital watermarking, the extraction algorithm fails. In this study, a watermarked fingerprint is implanted in an individual's facial picture.

Discrete Wavelet Transform

One of the most effective methods for signal representation is the wavelet transform. It is being used for signal processing, data compression, and picture processing. Only the frequency of information can be provided by a conventional Fourier transform. This processing results in the loss of temporal information. For example, in a musical score, we know not only the notes we want to play but also when to play them. In many applications, we need to know the temporal information and frequency at the same time. The image processing can be applied to this as matrix processing. Utilising time-frequency atoms with various time supports is necessary to analyse signals of drastically different sizes. The wavelet transform breaks down signals into wavelets, which are continuous functions that have been dilated and translated into highly redundant functions. The wavelet transforms in two dimensions differ slightly from those in one dimension.

It can be easily extended by merely multiplying the wavelet and one-dimensional scaling functions. In image processing, a two-dimensional wavelet transform is used. One two-dimensional scaling function is required for the two-dimensional wavelet transform. These functions measure the image's intensity variation in relation to various axes; some measure variations along columns, others along rows, and still others measure variations along diagonals. The scaling function yields an approximate result that is identical to the one-dimensional result. The analysis of extending the one-dimensional DWT to two dimensions is simple when the scaling function and wavelet functions are provided.

Any wavelet alteration for which the wavelets are analysed singly in numerical analysis and practise is referred to as a discrete wavelet change (DWT). The advantage of other wavelet changes over Fourier changes is that they capture both recurrence and area data (area on schedule). Discrete cosine change (DCT) and DWT both are applied for image pressure algorithms. If there should emerge an occurrence of DCT at a stage called quantization when sections of pressure truly happen, and the less important frequencies are disposed of, this causes a loss of information. The picture in the disintegration measure is then recovered using the primary frequencies that are still present. As a result, the reproduced image has been altered.

Because of this, we carry out DWT in this undertaking. Because of the energy compaction property of wavelets, DWT used for picture pressure produces a higher-quality image even when the pressure proportion is high. In DWT, the data is sent across a variety of low-pass and high-pass channels. The low-pass channel yield is the calculated coefficient, while the high-pass channel yield is point-by-point co-proficient. These outcomes were in seven subgroups where the surface provision of the unique mark picture was inserted.

Discrete wavelet transforms map data from the time domain to the wavelet domain. The result is a vector of the same size. Wavelet transforms are linear and they can be defined by matrices of dimension nxn if they are applied to inputs of size n. Depending on the boundary conditions, such matrices can be either orthogonal or "close" to orthogonal. When the matrix is orthogonal, the appropriate transform is a rotation in Rn in which the data is a point in Rn. The input image is passed through the low pass and high pass filters. The outputs obtained from these filters are then sampled. The output from the high pass filter is known as the detailed coefficient, and the output from the low pass filter is known as the approximate coefficient. This approach is employed in one-dimensional DWT as depicted in Fig. (**1**).

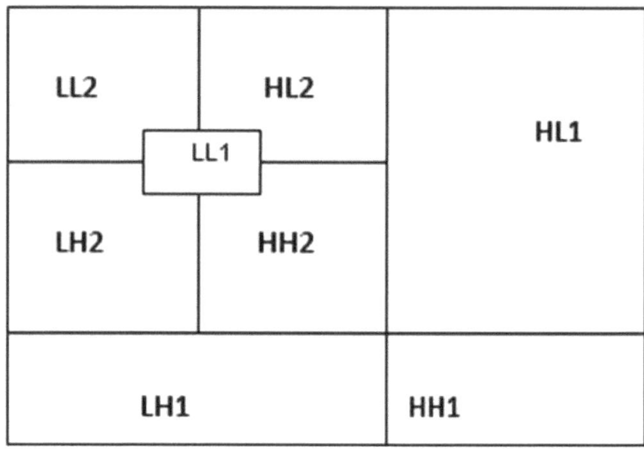

Fig. (1). Two level decomposition using DWT.

In two different approaches, a similar methodology is applied to the lines and sections of the two-dimensional DWT. The yields are then examined under both categories, resulting in the four coefficients LL, LH, HL, and HH. While the second alphabet symbolises the shift along the column, the first letter symbolises the alteration along the row. LH displays the signal as a high pass signal in the column and a low pass signal in the row. High pass signal is denoted by H, while low pass signal is denoted by L. Signal coding frequently uses the discrete wavelet change as a precondition for information pressure to address a discrete sign in a more repetitive structure. In order to use the wavelet change for image processing, the inspection and amalgamation channel banks must be represented in 2D. The 1D analysis filter bank is first applied to the image's columns in the 2D case, and then to its rows. If the image has N1 rows and N2 columns, we will have two sub-band images with N1/2 rows and N2 columns each after applying the 1D analysis filter bank to each column. We will also have four sub-band images with N1/2 rows and N2/2 columns each after applying the 1D analysis filter bank to each row of the two sub-band images. The diagram below provides an illustration of this. The four sub-band images are combined by the 2D synthesis filter bank to create the original N1 by N2 image. Similar to the 1D example, the low pass sub-band picture is iterated through the 2D analysis filter bank to execute the 2D discrete wavelet transform of a signal x. We may more precisely determine that it is a superior method in this situation since there are three sub-bands at each scale instead of the original one.

Inverse Discrete Wavelet Transform

IDWT is a cycle that allows fragments to be reassembled at the first indication without losing information. IDWT replicates a sign obtained from a disintegration-derived rough and point-by-point coefficient. IDWT differs from DWT in that it requires testing and sorting in a certain sequence. Upsampling, often referred to as interpolating, is the process of adding zeros to the signal in between samples. A single-level reconstruction is carried out using IDWT () using the provided coefficients. One kind of wavelet transform that is used is the inverse discrete wavelet transform (IDWT).

RESULTS

Extracting Face Textures

A surface with certain characteristics, such as roughness, granulation, and consistency, is used for the analysis of various types of images. For surface discernment, multi-scale preparation is shown using wavelet analysis. A key distinguishing feature of surface examination is the list of obtained capabilities that include energies of various scales. When a sign is degraded using a 1-dimensional wavelet change, the result is the altered sign. A 2-dimensional wavelet change is produced by applying a 1-dimensional change along the lines and sections. Using a Daubechies channel bank, the Discrete Wavelet Transform (DWT) is employed to build a pyramidal pattern of subgroups. The two-level decay has seven subgroups in it. The subgroups HH, HL, and LH contain the intricate diagonal, flat, and vertical patterns that make up the distinctive finger imprint image. The surface guide selects the areas whose size is different from the adjacent areas relative to a predefined edge. The watermarking calculation we suggest using these selected areas is based on these.

Embedding Fingerprint Watermark

Law enforcement authorization offices collect photographs of fingerprints, which they store in an information base with photographs of people's faces. In a dataset, the various information types are typically stored. Maintaining the chain of custody and the reliability of the fingerprints is essential, particularly when they are used as evidence in court. The distinctive mark is implanted in the face image to be verified and used as a pertinent watermark in this document. The black-and-white face image is divided into seven subgroups using a 2-level Discrete Wavelet Transform. The distinctive finger impression watermark photos are inserted with the wavelet coefficients of the face image that address the regions of the selected surface districts. Grayscale makes up the image of the unique finger impression. The determination of the watermarking subpicture is based on a few variables.

Because a big percentage of energy is accumulated in this band, changing the low recurrence subimage LL2 will cause serious degradation of the rebuilt picture. In the high-recurrence group, a portion of the data will be lost during separation and pressure. One way to overcome this data misfortune is to install data in all high-recurrence categories without a need (LH1, HL1, and HH1). LH2, HL2, and HH2 are the three mid-recurrence groups that are appropriate for implanting. For increased power, we implant the dim scale unique mark picture into the mid-recurrence group as shown in Fig. (**2**).

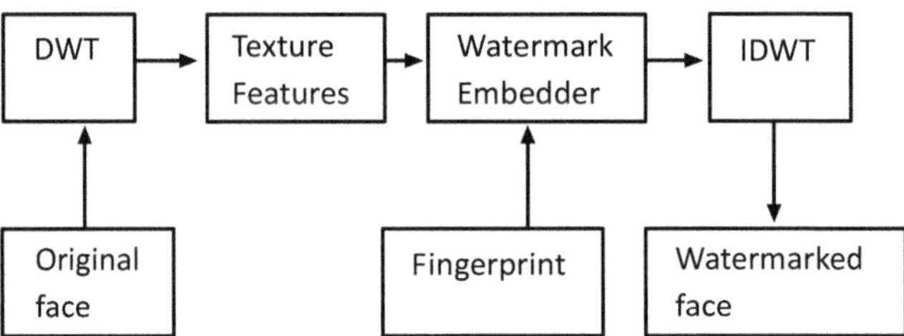

Fig. (2). Embedding watermark.

Extracting Fingerprint Watermark

A number of watermarked finger impressions have advantages. Changing the low recurrence subimage LL2 will seriously degrade the rebuilt picture because a significant portion of energy is accumulated in this band. Due to the high recurrence, some of the data will be lost during separation and pressure. The fact that the unique mark, the person's segment text data, and the face image do not need to be saved as separate data sets is one of the key benefits. Logic computerised watermarking enables simultaneous storage and recovery of all connected data. The extraction system functions the opposite way from the implanting system. Each of the six subpictures has its selected surface district determined, and the watermarks are taken from the most shortened request of the whole number of the comparing sub-pictures. The spatial excess in the mid-recurrence channel during the insertion of the distinctive finger imprint watermark guarantees robust extraction when two or more of the three qualities are similar. In a similar way, the finger impression image is removed from the lower recurrence channel. Using the extracted watermarks as shown in Fig. (**3**), it is possible to examine the integrity of the finger impression for any potential alterations.

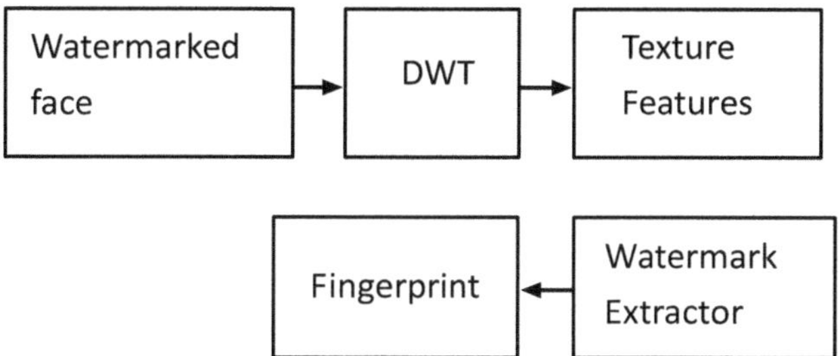

Fig. (3). Extraction of watermark.

Integrity Verification of Watermark

There are some advantages of the watermarked face. The fact that the face and facial image need not be stored in separate information bases is one of the main advantages. The logical computerised watermarking enables the simultaneous storage and recovery of all connected information. Recovery of the distinctive finger impression also helps to distinguish the evidence of an individual. The text and facial image will be deleted using this tactic. The implanting system's opposite is the extraction cycle. Currently, the request for the pieces to be removed is made using the same mystery key that was used during implanting. The chosen surface area is acquired for each of the six subpictures, and the watermarks are removed from the subpicture with the lowest request number for comparison. The spatial repetition provided in the high recurrence channel during message watermark installation ensures strong extraction when two or more of the three attributes are comparable. The unique mark picture is extracted from the lower recurrence channel using a similar technique.

For extraction, neither the first face image nor the first watermark image is required. The photos of the extracted finger impression are of acceptable quality and closely resemble the original photos. Using two distinct types of measurements, we quantitatively determine the degree of comparability between the original watermark image and the removed watermark images. The pixel-based resemblance between the initial and revised photographs is determined by the relationship between them. Both the human visual framework approach and the pixel-based methodology demonstrate a significant degree of connection between the photos in the comparability esteems. When exposed to various attacks, the watermarked face was strong and tough. The impact of these assaults on the distinct unique mark photographs for differentiating proof purposes is then determined as shown in Fig. (4).

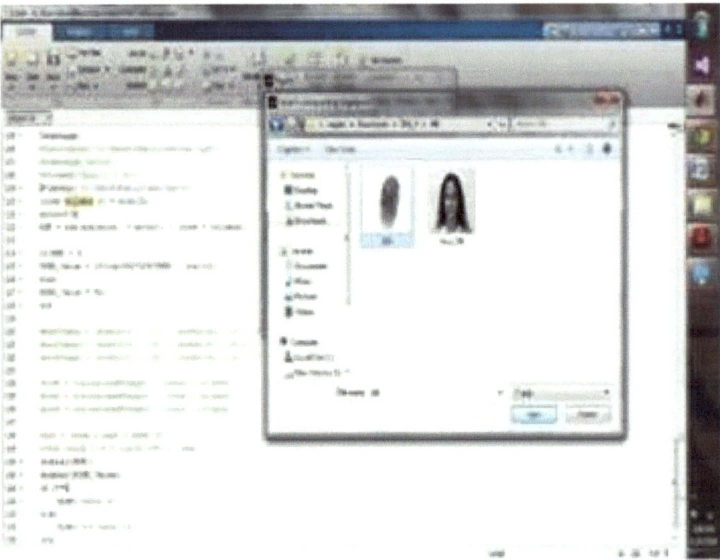

Fig. (4). Face and fingerprint input image.

PERFORMANCE EVALUATION

Peak Signal-to-Noise Ratio

The ratio of a sign's maximal theoretical force to the effect of undermining commotion is known as the peak signal-to-noise ratio. Human impression of reproduction quality is measured by PSNR.

The ratio of a sign's maximum potential value to the mutilating disturbance that affects the way it is shown is known as the sign to clamour proportion (PSNR). The PSNR is transmitted as far as the logarithmic decibel scale since many indications have an extraordinarily broad forceful reach (the ratio between the greatest and smallest possible upsides of an alterable quantity). It might be emotionally taxing to focus on the visual aspects of advanced pictures or to enhance them. Fig. (5) illustrates how, while claiming that one approach provides a higher quality, the image may differ depending on the person. For this reason, it's crucial to implement quantitative/observational measurements to assess how picture improvement calculations affect image quality. A comparable set of tests, images, and unique picture upgrading algorithms may be compared carefully to see if a certain calculation produces superior results. The maximum signal-to-commotion ratio is the measurement under consideration. The algorithm is depicted in Fig. (6) and if we can demonstrate that a calculation or set of calculations can improve a degraded realised picture to all take after the first more intently, then at that point we can all more precisely infer that it is a superior algorithm.

Fig. (5). Watermarked face image.

Fig. (6). Recovered watermarked image.

Mean Square Error (MSE)

The computerised image preparation approach includes a component that uses mean square errors to check for errors. Comparing packing an image to compressing basic double information, is completely different. It is obvious that pressure projects may be used to pack photographs, but the results are not precisely optimal. This is because encoders specifically designed for images might benefit from particular factual qualities that pictures contain. Additionally, some of the picture's more delicate details may be lost in order to save a little bit of transmission speed or space. This also means that lossy pressure techniques may be used in this area. Lossless pressure comprises information that is compacting and will be a precise replica of the original information upon depressurization. When executable archives and other types of double information are packed, for example, this is the scenario. When depressed, they should mostly be reproduced. However, it's not necessary to recreate the images "exactly". As long as the difference between the first and condensed picture is typical, an estimate of the first picture will work for the majority of applications. Two of the fundamental metrics used to examine the different picture pressure approaches (PSNR) are the Mean Square Error (MSE) and the Peak Signal-to-Noise Ratio. The PSNR is the total squared error between the packed and first picture, whereas the MSE is the percentage of the pinnacle error. The following are the mathematical formulae for both.

$$MSE = 1 \div MN\big(sum(y = 1ton)sum(x = 1ton)\big)[I(x,y) -$$
$$I'(x,y)^\wedge 2)PSNR = 20 * \log 10\left(\frac{255}{sqrt\,(MSE)}\right) \tag{1}$$

Where I(x,y) represents the first picture, I'(x,y) represents the approximated adaption (which is actually the de-pressurized picture), and M,N represent the picture elements. A lower MSE incentive implies fewer errors, and as the opposing relationship between MSE and PSNR shows, this translates to a high PSNR value. A greater PSNR value is consistently acceptable because it signifies a higher signal-to-Noise ratio. The 'signal' here is the initial image, and the 'commotion' is the recreating error. As a result, if you come across a pressure plot with a lower MSE (but a high PSNR), you can deduce that it is a better one.

CONCLUSION

In order to conceal a finger imprint image from the matching face image, this study proposes an effective multimodal biometric picture watermarking technique. The method was created to provide more secure biometric information verification while increasing recognition accuracy and decreasing transmission speed. The unique mark image was concealed in the cover face picture's DWT coefficients using the newly introduced calculation. The suggested method's presentation has been evaluated and contrasted with a variety of current multimodal biometric watermarking techniques and DWT-based methods. Peak signal noise ratio (PSNR) and mean squared error (MSE), two measurements with better advantages when the amount of hidden information is high, demonstrate the sufficiency of the proposed method while maintaining the intangibility and excellent watermarked picture quality in any case, and in holding the practical usefulness of both the biometric picture in any case.

REFERENCES

[1]　S. Sasi, K.C. Tamhane, and M.B. Rajappa, "Multimodal biometric digital watermarking on immigrant visas for homeland security", In: *Biometric Technology for Human Identification.* vol. Vol. 5404. SPIE, 2004, pp. 425-435.
[http://dx.doi.org/10.1117/12.537112]

[2]　K. Stefan, and AP, P. Fabien, *Information hiding techniques for steganography and digital watermarking.* Artech House Books, 2000.
[http://dx.doi.org/10.1201/1079/43263.28.6.20001201/30373.5]

[3]　U. Uludag, B. Günsel, and M. Ballan, "A Spatial Method for Watermarking of Fingerprint Images", *Conference: Pattern Recognition in Information Systems, Proceedings of the 1st International Workshop on Pattern Recognition in Information Systems, PRIS 2001, In conjunction with ICEIS 2001,* Setúbal, Portugal,July 6-7, 2001.

[4]　U. Uludag, and U. Uludag, "Hiding biometric data", *IEEE Trans. Pattern Anal. Mach. Intell.,* vol. 25, no. 11, pp. 1494-1498, 2003.
[http://dx.doi.org/10.1109/TPAMI.2003.1240122]

[5]　D. Moon, T. Kim, S. Jung, Y. Chung, K. Moon, D. Ahn, and S.K. Kim, "Performance evaluation of watermarking techniques for secure multimodal biometric systems", *Computational Intelligence and Security. CIS 2005. Lecture Notes in Computer Science,* Berlin, Heidelberg, vol.3802, pp.635-642, 2005.
[http://dx.doi.org/10.1007/11596981_94]

[6]　T. Hoang, D. Tran, D. Sharma, T. Le, and B.H. Le, "Priority watermarking-based face-fingerprint authentication system", *2009 International Conference on Information and Multimedia Technology.* Jeju, Korea (South), 16-18 December,pp.235-238, 2009.
[http://dx.doi.org/10.1109/ICIMT.2009.44]

[7]　A. Ross, and A.K. Jain, "Multimodal biometrics: An overview", *2004 12ᵗʰ European signal processing conference.* Vienna, Austria, 06-10 September,pp.1221-1224, 2004.

[8]　A. Ross, and A. Jain, "Information fusion in biometrics", *Pattern Recognit. Lett.,* vol. 24, no. 13, pp. 2115-2125, 2003.
[http://dx.doi.org/10.1016/S0167-8655(03)00079-5]

[9]　L. Hong, A.K. Jain, and S. Pankanti, "Can multibiometrics improve performance?", In: *Proceedings AutoID.* vol. 99. Citeseer, 1999, pp. 59-64.

[10]　A.S. Tolba, and A.N. Abu-Rezq, "Combined classifiers for invariant face recognition", *Pattern Anal. Appl.,* vol. 3, no. 4, pp. 289-302, 2000.
[http://dx.doi.org/10.1007/s100440070001]

[11]　A.K. Jain, and Jianjiang Feng, "Latent fingerprint matching", *IEEE Trans. Pattern Anal. Mach. Intell.,* vol. 33, no. 1, pp. 88-100, 2011.
[http://dx.doi.org/10.1109/TPAMI.2010.59] [PMID: 21088321]

[12]   A.A. Ross, K. Nandakumar, and A.K. Jain, *Handbook of multibiometrics.* vol. 6. Springer Science & Business Media, 2006.

[13]   Y. Wang, T. Tan, and A.K. Jain, "Combining face and iris biometrics for identity verification", *International conference on Audio-and video-based biometric person authentication,* Springer, Berlin, Heidelberg, pp.805-813, 2003.
[http://dx.doi.org/10.1007/3-540-44887-X_93]

[14]   S.C. Dass, K. Nandakumar, and A.K. Jain, "A principled approach to score level fusion in multimodal biometric systems", *International conference on audio-and video-based biometric person authentication,* Springer, Berlin, Heidelberg, pp.1049-1058, 2003.
[http://dx.doi.org/10.1007/11527923_109]

[15]   U. Uludag, and U. Uludag, "Hiding biometric data", *IEEE Trans. Pattern Anal. Mach. Intell.,* vol. 25, no. 11, pp. 1494-1498, 2003.
[http://dx.doi.org/10.1109/TPAMI.2003.1240122]

CHAPTER 3

Underwater AUV Localization with Optimal Cardinal Selection Using Dynamic Positioning Parameters

Prashanth N.A.[1] and **Prasanth Venkatareddy**[2,*]

[1] *Department of Electrical and Electronics, BMS Institute of Technology and Management, Bangalore-560064, India*

[2] *Department of Electrical and Electronics, Nitte Meenakshi Institute of Technology, Bangalore-560064, India*

Abstract: Today, underwater communication has become a hot issue in research on both undersea and deep-sea navigation, as well as in autonomous underwater vehicle management, and acoustic communication has been accounted for due to its flexibility and lower degree of attenuation. However, owing to influencing elements such as channel time changing circumstances, bandwidth measurements, longer propagations delay and the greatest degree of Doppler spread, pressure conditions, and salinity level, establishing acoustic communication in real-time is much more difficult.

With a new monitoring era of global physical entities, a new agent-based multipath routing protocol has been proposed in this work including underwater sensor nodes and underwater gateways with an autonomous underwater vehicle (AUV). The clustering head in the impacted region of sensor nodes will gather and aggregate data using mobile agent-initiated routing algorithms for identifying numerous pathways, as well as parameters including hope counting, delay propagation, nodal energy, and channel quality. In this paper, an agent-based dynamic AUV traversal method is developed for increasing the network's dependability and connection while reorienting the AUV's movement direction.

Keywords: Challenges, Routing, Radio waves, Sensors, Underwater communication.

INTRODUCTION

There are many challenges existing in underwater sensor networks like GPS absence, node deployment in the 3D method, and water currents with mobility [1, 2] along with the biggest challenge of using radio waves with feasibility for

---

* **Corresponding author Prasanth Venkatareddy:** Department of Electrical and Electronics, Nitte Meenakshi Institute of Technology, Bangalore-560064, India; E-mail: prasanth.v@nmit.ac.in

S. Kannadhasan, R. Nagarajan, Alagar Karthick, K.K. Saravanan & Kaushik Pal (Eds.)

underwater communication. Though the radio waves propagate through shorter distances, they will get faded inside water due to larger absorption thus making the node design methodology more complicated as the large waves will require the larger antenna. Hence, acoustic waves are being utilized for longer distance underwater communication but, still they also have challenges like high noise, the highest bit rate error, the Doppler effect, and the highest propagation delays [3, 4] and it is a must to consider these characteristics while designing the sensor network [5] for underwater communication and the general architecture of it is shown below in Fig. (1).

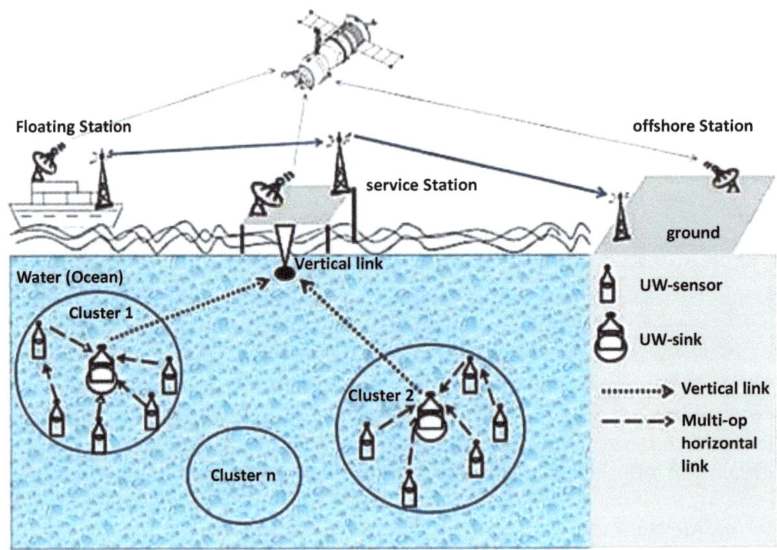

Fig. (1). Underwater sensor network architecture [10].

The architecture consists of either the static sensor nodes or the mobile sensor nodes being equipped with the acoustic transceivers which will be deployed in underwater. Depending upon the sink nodes, the routing of sensor nodes will occur as relays in the lower level of the depth whereas the autonomous underwater vehicles (AUV) will be employed directly to receive data from the sensors, and it requires an efficient path for routing.

In general, while the designing process of UWSNs will be carried out, routing protocols will be playing an efficient role thereby determining the routing data paths along the paths from sensor nodes to interpreted stations. Robust protocols are selected to achieve optimal performance against harsh channel conditions by dealing with energy constraints efficiently with the highest SINR, higher level of end-to-end delays, and the node mobility [6-9]. Though many surveys will not be focusing on the recent work, some of them do discuss the advantages and the disadvantages of the routing protocols with a better understanding of the protocols [8, 9].

ISSUES TO BE ADDRESSED BY ROUTING PROTOCOL

The List of Issues Addressed by the Routing Protocols is as Follows

i. Estimating the Location – This is the challenging part due to GPS unavailability underwater by using the routing protocols incurred owing to localization.

ii. Required Location - The required geographical location will be indicated and this is the main issue to be addressed by the proposed scheme.

iii. Constraints of energy- Energy constraints of UWSN are considered as the batteries will neither be recharged nor replaced [10] by considering the energy conversation.

iv. Mobility- Underwater sensor node moment may interrupt end-to-end paths by creating void connectivity from the source to the sink with free movements of nodes.

v. Connectivity avoidance-If the node lies on the route of packets, connectivity voids will be created owing to the malfunctions or the energy drainage, there avoiding the voids with the balanced loads and delivering the packets to the destination nodes successfully.

vi. Deployment in 3D - Due to the additional third dimensions,3D deployment will be quite challenging as the sensor nodes can rely on data to sink forward multi-hop paths using the current transmission.

vii. Clustering- It is used to decide whether this is a cluster-based scheme or not with lower communication overhead under a small contention domain with lower power of transmission.

viii. Selection criteria of the next hop-This is done to determine the proposed scheme's capability to address routine issues such as residual energy by considering the SNR or the link condition.

Challenges in Underwater Communications

Considering the challenges involved in underwater communication, a UWSN routing protocol will be designed, and the related necessities of the challenges are mentioned below.

Noise in Underwater

The communication quality might be affected due to the noise present in the underwater and hence, soundless paths will be selected for routing protocols. There are four components of ambient noise underwater *viz.*, thermal, shipping turbulence noise, and wave [11, 12]. In which the sum of the spectral power densities of these noises will be considered to calculate the power spectral density N of the ambient noise in dB as:

$$N = N_{sh} + N_{wv} + N_{tb} + N_{th} \qquad (1)$$

Here N_{tb}, N_{sh}, N_{th}, and N_{wv} represent spectral power densities of shipping turbulence, shipping, thermal noise and wave, respectively. They are modeled by the following equations (2) to (5) in which f denotes the frequency level in kHz, wind speed is denoted by w in m/s, whereas the shipping extent will be represented with s ϵ [0,1] in the water.

$$N_{sh} = 40 + 20(s - 0.5) + 26log(f) - 60\log(f + 0.03) \qquad (2)$$

$$N_{wv} = 50 + 7.5^{w} + 20log(f) - 40\log(f + 0.4) \qquad (3)$$

$$N_{tb} = 27 + 30log(f) \qquad (4)$$

$$N_{th} = -27 + log(f) \qquad (5)$$

By reducing the frequency below 20 Hz, the noise will be generated by the turbulence in the water ranging from 20Hz to 200Hz, with the wind blowing at the water surface, resulting in noise between 20 Hz to 200 Hz with the temperature generated thermal noise affecting at the frequencyof about 200 Hz.

Channel Attenuation

Due to the spreading loss and the absorption loss, tempering will be made in the underwater communications [13], thus decreasing the desired signal length of the routing, thereby receiving the signal at the required destination. The acoustic waves attenuation with frequency 'f' in kHz with source distance denoted as 'd' can be denoted and modeled with (D, F) as:

$$A(D, F) = A_0 d^k \alpha(f)^d \qquad (6)$$

In the above equation, the normalization constant is denoted by A_0, the coefficient of absorption is denoted by á, whereas the spreading factor is denoted by k. Aside, the value of K = 1 denotes the cylindrical spread geometry whereas the value of k=2 denotes the spherical spread geometry with the basic k=1.6.

Limited Bandwidth

Only the specific frequencies are allowed by the harsh underwater medium for carrying the information with restricted bandwidth on the acoustic design, thereby considering the routing protocols with limited frequencies for path selection and packet delivery to the destination, and it is found that [14] the transmission range of underwater is proportional to the bandwidth and all are shown below in Table 1.

Table 1. shows the relationship of convergence bandwidth with UWSNs.

Convergence	Range (km)	Band width (kHz)
Very long	100	< 1
Long	10-100	2-5
Medium	1 - 10	Almost 10
Short	0.1-1	20-50
Very Short	< 0.1	> 100

Based on the values of temperature, salinity, and depth of water, the acoustic waves will be varying in the underwater communication and are modeled with:

$$c = 1449 + 4.591T - 5.304 \; X \; 10^{-2}T^2 + 2.374 \; X \; 10^{-4}T^3 + 1.34(S - 35) + 1.63X10^{-2}D + 1.675X10^{-7}D + 1.025X10^{-2}T \; (S - 35) - 7.139X10^{-3}TD^3 \qquad (7)$$

In the above-mentioned equation, T represents temperature, D represents the sea depth, whereas S represents the salinity factor in parts per thousand under the temperature conditions of $0 \; ^\circ C < T \leq 30 \; ^\circ C$, $0 \; m \leq D \leq 8000m$ and range of salinity between 30 – 400 ppt. The RF waves will be inherently higher in propagation and the speed changes will affect the time of delivery during the routing protocols. Based on time-critical underwater applications, the routing protocol design will be changed.

2-S LOCALIZATION PROCESS FOR COMMUNICATION

The two-stage distributed localization is the two-phase time synchronization free localization algorithm (TP-TSFLA) and it has two stages *viz.*, the first stage and the second stage [15]. During the first stage, all the reference nodes will be localized whereas, during the second stage, those localized nodes will be acting as the reference nodes for the un-localized nodes left which will be reaching the localized node with increased range with a beacon. Due to those un-localized nodes, higher overheads will be incurred with a higher level of energy consumption. Without clustering, this technique can be called a simple two-stage localization, but it is not being used for localizing partitioned nodes as it assumes that it will be remaining un-localized if they receive proactive requests. If the node receives a localization beacon in response, it will be localized whereas if they are not able to receive a beacon, it will send a request again with increased power of transmission until it will be localized.

In addition, 2S decides that if any localized node request will receive the localization upon transmitting a beacon, a bigger communication process will be created when every node responds to the requests. Implementing this technique, the communication overheads for 20,40,60,80 and 100 nodes are shown in Fig. (**2**). This technique will not waste the underwater bandwidth thereby increasing the consumption of energy with an increased count of nodes from 20 to 100 numbers thereby with increasing overhead from 6 to 18 with scalability loss.

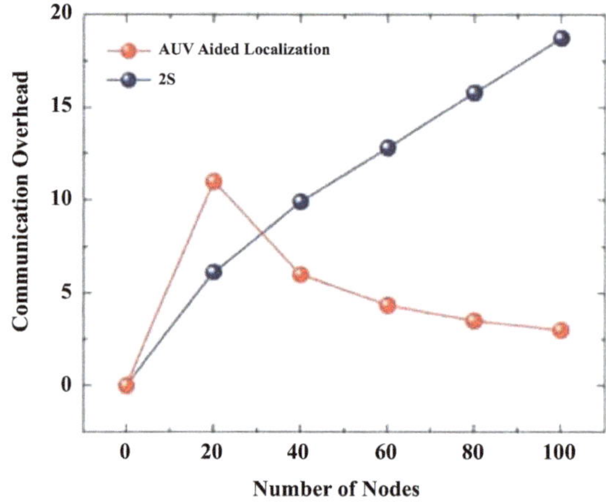

Fig. (2). Shows the communication overhead of 2S and AUV aided localization clustering.

Details and Mathematical Models for Network Environment Routing

The Environment of the Network

The network environment is discussed in this section and as shown in Fig. (3), a network having UW-GWs, sensor nodes, sinks, SH, AUVs, and son buoys is considered in which a set of N number is deployed with a layered approach as mentioned in a study with multiple SGs found in the first layer and the underwater nodes in the second layer with a node range depth, R in km to cover the depth of the ocean. Simultaneously, for creating a 2D grid over the surface, all the SGs having two interfaces will be placed uniformly with upper bounded deployments with 2R in which R refers to the SG cover range. The two interfaces are an acoustic interface and a wireless interface with a distance between them limited to 2R.

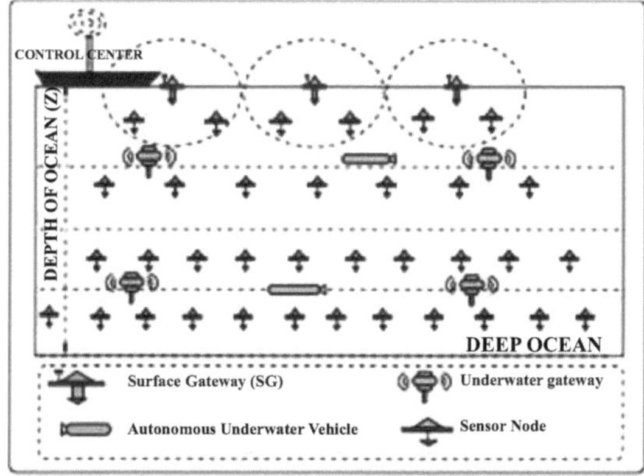

Fig. (3). shows underwater acoustic sensor network's architecture.

Mathematical Model

Mounted with sensors like salinity, pressure, temperature, and velocity sensor, the nodes will be programmed periodically to collect information and then compared with threshold values for deciding event occurrence with sensed parameters. If the sensed information is higher than the threshold, it will be divided into two groups namely, critical event and noncritical event. Here the velocity sensor-triggered event will be called a critical event, which is used for anti-submarine detection. Whereas the event triggered by the salinity pressure, temperature and velocity sensor is called a non-critical event, which is used for monitoring the environment as follows. Let's assume $D = D_{temp}, D_{saln}, D_{press}, D_{vel}$ with the information collected from respective salinity, pressure, temperature, and velocity sensors which will be compared with the threshold values $D_{tempth}, D_{salnth}, D_{pressth},$ and D_{velth}.

If the value of Di Is greater than the $Di_{th,}$ then it will be triggered with the 'I' value from the salinity level, value of pressure, and temperature at the sensors of the velocity. If the nodes are interfaced with sensors, the generalized equation will be represented as follows: Let D= D1, D2, D3…. DnCollected from sensors and the threshold values D1th, D2th, D3th….Dnth from the sensor's parameters for the event from ith sensor where i=1,2, 3…n. using equation 8, the event will be directed from the ith sensor nodeEdeti as shown below.

$$E_{det_i} = \begin{pmatrix} 1, & D_i > D_{i_{th}} \\ 0, & Otherwise \end{pmatrix} \tag{8}$$

Agencies

The AUVs consist of two agents like the static type agents and the mobile type gents in which the static agents are the vehicle blackboard (VBB) and the Vehicle Monitoring agent (VMOA), whereas the mobile agents the vehicle data receiving agent (VDRA), Vehicle path discovery agent (VPDA), Vehicle traversal agent (VTA) and the vehicle Data Transmission Agent (VDTA) as shown in Fig. (4).

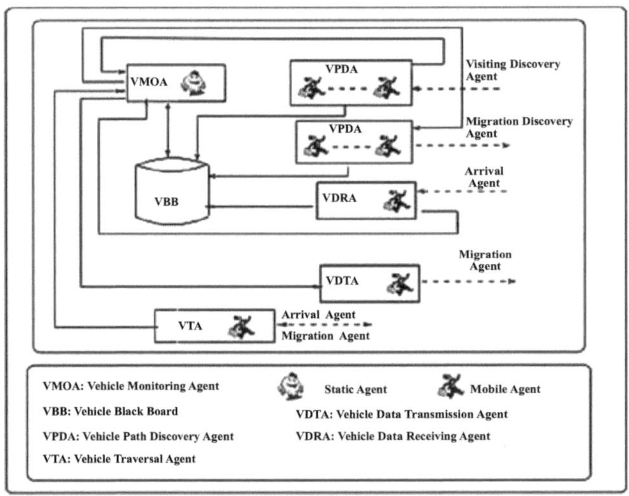

Fig. (4). Shows the agency details of Vehicle.

The node agency of the vehicle plays a vital role in partitioning the network due to dead node conditions and the failures of the link and with the help of AUV, the data will be transmitted even though the nodes are dead nodes or due to mistaken nodes and the functionality of agents as described below.

The functionalities of agents are:

i. VBB Neighbors will be found and updated by VMOA to NMOA by reorienting the path movements.

ii. It is found that VBB Function is like NBB.

iii. Either from vehicle to vehicle or from vehicle to node, VPDA will be established with movement orientation by collecting data from the network.

iv. From UG-GW, information will be collected by VTA and then fed to VMOA.

The collected information will be disseminated by VDTA.

v. Information will be collected by VDRA and stored in VBB

AGENT-BASED ROUTE DISCOVERY PROCESS

A CH will be selected among cluster members and will aggregate the material during route discovery. The beginning party is set up, the NMA is initiated on triggering the event, and the CH is chosen. When an application request reaches the node's NMA or an AUV's VMA, the generation process of route discovery agents is started and propagated to all of the node's neighbours. Through message exchange, the NBB gives the neighbour information from the NMOA. The NBB is routinely updated by the NMOA *via* an independent procedure.

The NPDA is in charge of finding routes and sending path data to the SG. The information will be returned to the source node for continuous data transmission after being obtained from several NPDAs thanks to the assistance of many clones of any NPDA. Members of the routing agency that aids in establishing the various pathways include NPDA, NMA, and NMOA. The steps involved in route finding are shown below. 1) The NMA obtains the list of neighbours' data from the NMOA. 2) Create as many NPDA clones as there are neighbour nodes. 3) A route package will be prepared and sent through NPDAs to each of your neighbours. 4) The request packet contains the source and destination addresses as well as the sequence ID field.

The RRQ packet contains the following data: ID, SA, DA, RI, RDF, and Seq No. The destination address, the root discovery flag (RDF), and the packet identification are among the different fields in an RRQ packet. 5) The routing table will be supplied to the source node, which will create it before broadcasting the RRQ to its neighbours, when the destination node gets an RRQ packet. The route reply message (RRS) is what the intermediary nodes send. 6) The RRS packet contains the following data: RDF=RRS=ID, SA, DA, Nh, Lq, Nq, Ne, RI,

RDF=RRS=ID, SA, DA, Nh, Lq, Nq, Ne, RDF. Numerous node metrics, including link quality (Lq), queue length (Nq), node energy (Ne), and hop count (Nh), are supplied as an attachment to the RRQ message in order to examine and assess the quality of neighbouring nodes. 7) The node will get a large number of packets from various pathways on the vector routing table scenarios.

Loops will be avoided during the route discovery phase, and if the receiving node is an RRQ packet, an individual ID will be produced based on the ID sequence of the RRQ packets. The gateway nodes update the routing database with the resultant gateway IDs and information on the hop count. Later, the AUVs' traversal paths will be created using this data. 8) The routing table NB, which is used to unicast the RSS message to the source node, is updated when an RRS message is received. 9) The SGMA discards the RRQ packet if it gets multiple copies of the same packet after creating the RRS packet. 10) After the RRS is received, nodes to RDF node sets will be produced and NMA will be transferred again to the next node in the RSS packet field's RI field. 11) When RSS reaches the source node, the data is handled in accordance with the routing table without the need for broadcasting messages from various nodes that have been formed as SG.

SIMULATION

Depending on the clustering method and routing in various studies, the recommended agent-based approach (ABA) will be compared. The technique of clustering will be compared with the ratio of packet delivery, consumption range, and delay value, as indicated in a study. The non-agent-based routing method (NABRA) and non-agent-based clustering approach (NABCA) are called from the simulation findings in the works and NS3 was utilised to acquire the results.

Simulation Model

In a study, a scenario with sensor nodes was evaluated with AUVs, UW-GWs, and numerous SGs placed inside the 500 m grid with a depth of 1km to 8km. Table **2** lists the many parameters that were used in the simulation setup. Some of the parameters were derived from real-time implementation traces. Sea Glider and Remus models for AUVs are used in the simulation.

Table 2. Shows the simulation parameters.

Parameters	Values	Parameters	Values
Number of Nodes	20 to 100	Pause time	5s
Range of sensor Node	50m,100m,150m	Call back	0m (water surface)

(Table 2) cont.....

Parameters	Values	Parameters	Values
Energy of Sensor Node	5 Joules	Slot time	0.2 s
Submerged depth	Vertical upto 100m	Bandwidth	4000Hz
Transmission Power	Varied from 2mW to 8mW	Default step time	10
Mobility Model	Random Waypoint	Arrival data rate (Poisson process)	0.002-0.1 pkts/msec
Vertical speed	0.2- 0.4 m/s	Service rate	0.001pkts/msec
Moment Duration	20s	Payload size	100B, 512B,1000B
Queue Size	50 bytes	Carrier frequency	256 Khz

Simulation Procedure

The simulation approach involves the following steps: 1) The creation of a UWASN environment with a sensor node, together with several AUVs, SGs, and UW-GWs, 2) The process of event detection, 3) The implementation of the suggested model, and 4) The assessment of the system's nature and performance.

The recommended routing algorithm's performance may be measured using the performance metrics listed below.

1) Clustering time: This is the total time required to create a cluster when an event takes place.

2) Clustering energy: This is the total energy expended in the clustering process.

3) Aggregation Time: This is the total time required to distinguish between the date and the cluster head and event node.

4) Aggregation energy: This refers to the total amount of energy used by the cluster's nodes to aggregate data.

5) The source and destination nodes' end-to-end delay will be estimated as a unit of distance.

6) Packet delivery ratio (PDR) is the percentage of packets that are successfully delivered from the source node to the destination node.

7) Energy consumption shows how much power is used by network nodes to send data to the target node, per pit.

8) The quantity of data points detected per unit of time is referred to as traffic.

RESULTS

Time and Energy of the Clustering Process

The recommendation ABA, as shown in Fig. (5), requires less time to construct a cluster since the CH is selected according to the channel characteristics, energy value, and depth level. The sensor that is closest to the seabed has its notes set to CH, which shortens the time it takes for clusters to form and speeds up data transmission. Although the rate at which nodes exchange messages is crucial, NABCA does not take this into account, because the NMA already knows whom its neighbors are in the planned ABA at the NBB., Fig. (5) shows the time for cluster formation *versus* nodes in msec.

The cluster nodes and sensor node locations will determine how long will it take to establish a cluster. Less time will be required by the clustering algorithm for nodes closer to the seabed (more than one km down to depths of 8 km) than for nodes closer to the sea surface (less than 1 km). As the number of nodes within the cluster rises, as shown in Fig. (5), so does the energy needed to build the cluster. This is true since many transmissions and receptions need energy. However, the suggested solution uses less energy than NABCA since the agents, who are endowed with the ability to make decisions, did not produce acknowledgment packets.

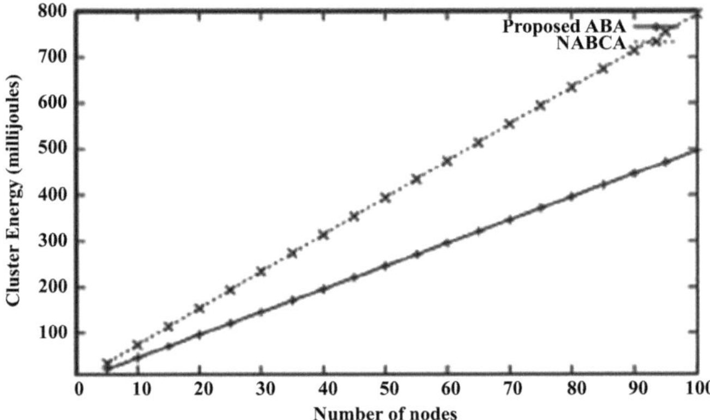

Fig. (5). Shows the energy of clustering *versus* nodes in milli joules.

ENERGY CONSUMPTION

The energy required to create a route and transfer from the event node CH to the sink node is shown in Fig. (6). Less energy is used by the proposed technique than by NABRA, and it is destroyed if the visited node does not measure the node quality. Also, retransmission will be lowered if the visited node does not match the quality of the neighbouring node. The energy consumption of the ABA and NABRA is shown in Fig. (7) as traffic and the number of nodes grow.

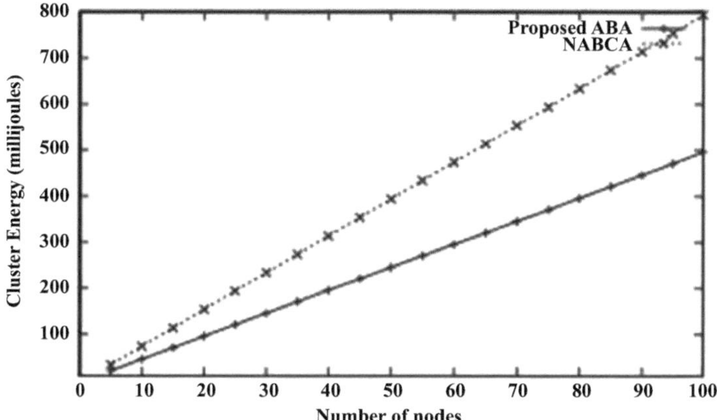

Fig. (6). Shows the energy consumption *versus* nodes in millijoules.

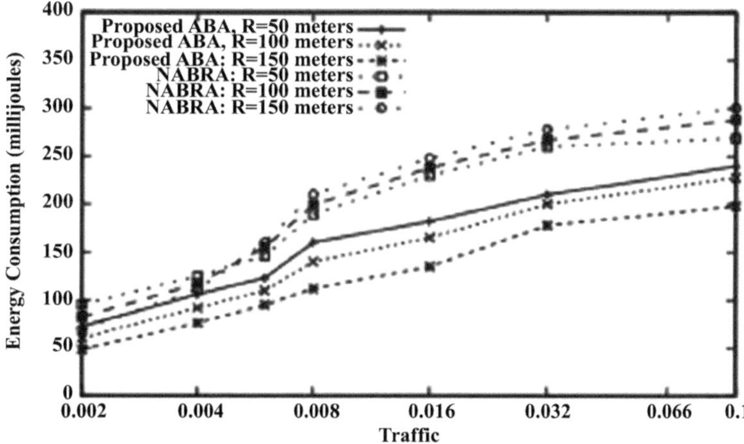

Fig. (7). Shows the consumption of energy consumption *versus* traffic in millijoules.

The recommended solution uses less energy because of the quality of the data transfer path's group link nodes, queue length, node energy value, and low number of hops. Additionally, when the range nodes increase, fewer relaying nodes are needed to reach SG, which reduce energy consumption.

PACKET DELIVERY RATIO

Fig. (8) shows the packet delivery ratio for the proposed technique and NABRA with rising traffic and for various payloads.

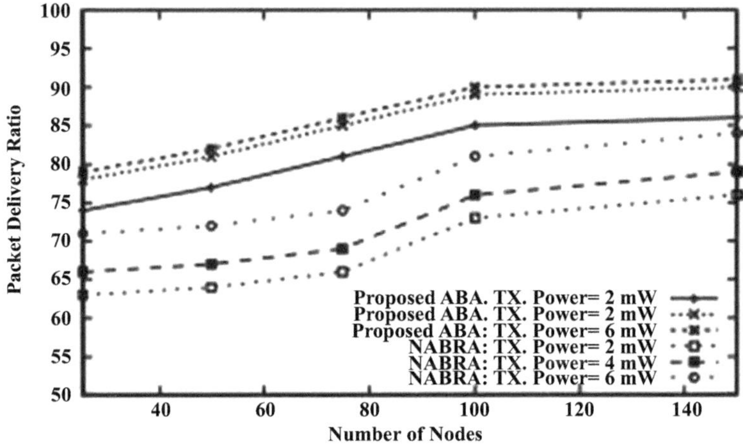

Fig. (8). Shows the ratio of packet delivery to the nodes.

The suggested approach has a higher PDR because relay nodes are selected based on more than just link quality; in addition, it will be done on energy and depth, for faster and more efficient data transmission. Furthermore, NABRA's link quality will be derived based on reliable data transmission, which can change with network operating time and topology changes.

CONCLUSION

Here, a standard agent-based routing strategy with a variety of agencies at the sensor node is presented. Uw-Gws, AUVs, and SGs are tools for efficiently forming clusters, analysing data, locating neighbours, aggregating data, and setting up paths with priority. Relay nodes are chosen using an assessment of DOA based on the connection quality. The relay nodes will also be chosen in two stages, the route maintenance phase and the road discovery phase, depending on the hop count, node depth, node queue, and node energy. The partitions of the network will be reoriented in the AUV direction based on the calculation of DOA, and the failures of the nodes will be taken into account for covering the partitioned region of the network. According to the simulation findings of the proposed method, the performance is higher when the energy consumption is combined with improved clustering timing, end-to-end delay, aggressive data timing, and the energy spent to attend this. Future proposals for better agent-based routing will include increased privacy protections and dependability.

REFERENCES

[1]   E. Felemban, Shaikh FK, Qureshi UM, Sheikh AA., and Qaisar SB., "International Journal of Distributed Sensor Networks", *Under water sensor network applications: A comprehensive survey,*

vol. 11, no. 11, p. 896832, 2015.

[2] Z Wang, X Feng, H Qin, H Guo, and G Han, "An AUV-aided routing protocol based on dynamic gatewaynodes for underwater wireless sensor networks", *Journal of Internet Technology.*, vol. 18, no. 2, pp. 333-43, 2017.

[3] N. Goyal, M Dave, and AK Verma, "Protocol stack of under water wireless sensor network", *classical approaches and new trends. Wireless Personal Communications.*, vol. 104, no. 3, pp. 995-1022, 2019.

[4] A Khan, I Ali, A Ghani, N Khan, M Alsaqer, AU Rahman, and H Mahmood, "Routing protocols forunderwater wireless sensor networks: Taxonomy, research challenges, routing strategies and futuredirections", *Sensors,* vol. 18, no. 5, p. 1619, 2018.

[5] Han G., Long X., Zhu C., Guizani M., and Zhang W., "Ahigh-availability data collection scheme based on multi-AUVs forunder water sensor networks", *IEEE Transactions on Mobile Computing.*, vol. 19, no. 5, pp. 1010-22, Mar 27, 2019.

[6] Nguyen NT, Le TT, Nguyen HH, and Voznak M., "Energy-efficient clustering multi-hoprouting protocol in a UWSN", *Sensors.*, vol. 21, no. 02, p. 627, Jan 2021.

[7] S. Karim, F.K. Shaikh, K. Aurangzeb, B.S. Chowdhry, and Al hussein, "M. Anchor nodes assisted cluster-basedrouting protocol for reliable data transfer in underwater wireless sensor networks", *IEEE Access,* vol. 9, pp. 36730-36747, 2021.

[8] O.A. Mahdi, A.B. Ghazi, and Y.R. Al-Mayouf, "Void-hole aware and reliable data forwarding strategy forunder water wireless sensor networks", *Journal of Intelligent Systems.*, vol. 30, no. 1, pp. 564-77, Jan 1, 2021.

[9] Jan S., Yafi E., Hafeez A., Khatana HW., Hussain S., Akhtar R., and Wadud Z., "Investigating Master–Slave Architecturefor Underwater Wireless SensorNetwork", *Sensors.*, vol. 21, no. 9, p. 3000, 2021.

[10] Islam T, and Lee YK, "A cluster-based localization scheme with partition handling for mobile underwateracousticsensor networks", *Sensors.*, vol. 19, no. 5, p. 1039, 2019.

[11] Islam T, and Lee YK, "A comprehensive survey of recent routing protocols for underwater acoustic sensor networks", *Sensors.*, vol. 19, no. 19, p. 4256, 2019.

[12] Bharamagoudra MR, Manvi SS, and Gonen B., "Event driven energy depth and channel aware routing forunder water acoustic sensor networks: Agent oriented clustering-based approach", *Computers & Electrical Engineering,* vol. 1, no. 58, pp. 1-9, 2017.

[13] I. García-Magariño, R. Lacuesta, and J. Lloret, "ABS-FishCount: An agent-based simulator of under water sensors for measuring the number of fish", *Sensors,* vol. 17, no. 11, p. 2606, 2017.

[14] Robinson YH, Vimal S, Julie EG, Khari M, Expósito-Izquierdo C, and Martínez J, "Hybrid optimizationrouting management for autonomous underwater vehicle in the internet of underwater things", *Earth Sci. Inform.,* 2021.

[15] N Li, JF Martínez, MenesesChaus JM, and Eckert M., "A survey on underwater acoustic sensor networkroutingprotocols", *Sensor,* 2016.

CHAPTER 4

Detection of COVID-19 Pandemic Face Mask Using ConvNet in Busy Environments

Veluchamy S.[1], Rajeesh Kumar N.V.[2], Srinivasan P.[3], Nandhakumar A.[4,*] and K.G. Parthiban[4]

[1] *Department of Electronics and Communication Engineering, Sri Venkadeswara College of Engineering and Technology, Coimbatore, India*

[2] *Computer Science and Engineering, Amrita College of Engineering and Technology, Coimbatore, India*

[3] *Department of Electronics and Communication Engineering, Amrita College of Engineering and Technology, Coimbatore, India*

[4] *Department of Electronics and Communication Engineering, Dhaanish Ahmed Institute of Technology, Coimbatore, India*

Abstract: The number of people using face masks has increased on public transportation, retail outlets, and at the workplace. All municipal entrances, workplaces, malls, schools, and hospital gates must have temperature and mask checks in order for people to enter. The paper's goal is to find someone who isn't wearing a face mask in order to control COVID-19. ConvNets may be used to recognize and classify images. The model depends on ConvNot to assess whether or not someone is wearing a mask. It is possible to identify an image's face by utilizing a face identification algorithm. These faces are then processed using Conv Net face mask detection. If the model is able to extract patterns and characteristics from photographs, it will be categorized as either "Mask" or "No Mask". With an accuracy rate of 99.85 percent, Mobile Net V2 is the most accurate in regard to training data. MobilenetV2 correctly identifies the mask in "Mask" or "No Mask" video transmissions.

Keywords: Conv Net, Covid-19, Face Mask Detection, MobilenetV2, Open CV.

INTRODUCTION

During the worst public health outbreak in recorded history, the COVID-19 virus sparked it. Wearing a face mask is the most effective way to prevent the virus from spreading. Droplets from the infected person could be the potential medium to propagate the infection. One can protect themselves from the disease by wearing a mask as illustrated in Fig. (**1**). One must wear a mask that totally covers

---

* **Corresponding author Nandhakumar A.:** Department of Electronics and Communication Engineering, Dhaanish Ahmed institute of technology, Coimbatore, India; E-mail: nandhakumar3107@gmail.com

S. Kannadhasan, R. Nagarajan, Alagar Karthick, K.K. Saravanan & Kaushik Pal (Eds.)

one's nose and mouth in order to protect them. Incorrect use of the mask could result in the transmission of disease and inadequate protection. Face-mask recognizers can be used in public settings to keep looking on the crowd and identify people who are or aren't wearing masks correctly. One possible purpose for this is to spread the word about proper mask usage and to help others learn how to do it themselves. In regard to the public use of face masks, Covid-19 has witnessed an uptick in popularity. Face identification is becoming increasingly difficult as the number of people using Covid-19-resistant masks rises. Wearing masks and avoiding eye contact have become increasingly trendy [2].

COVID-19 prompted governments around the world to make face masks and other protective gear essential for the public. People are now wearing face masks more frequently in settings where people come into close proximity, such as public transit, shopping malls, and the workplace. To avoid dizziness caused by carbon dioxide retention rebreathing, long-term use of face masks should be avoided [3]. The precautionary principle has been in effect since March of this year. One of the most recent safety measures was the use of surgical-style masks to cover our mouths and nostrils [4]. The current COVID-19 outbreak has resulted in an unprecedented loss of human life and severe post-disease repercussions for people of all ages, living situations, health statuses, and other distinguishing traits [5]. The term "single-use mask" refers to a mask that is intended to be worn only once before being discarded as shown in Fig. (**1**). Medical professionals and the general public alike favour single-use face masks for their high filtration capacity, lightweight, low cost, ease of use, breathability, and ease of disposal [6].

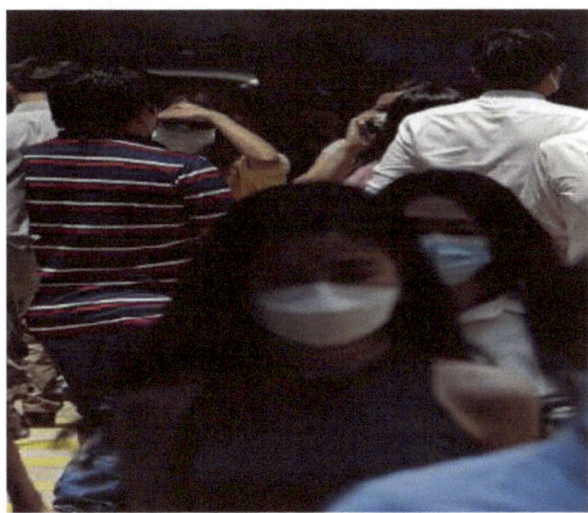

Fig. (1). Using a mask in a crowded environment.

Thrown into bushes, water bodies on the street, and mixed with regular trash, face masks are also seen abandoned at beaches, in waiting areas, dust/toilet bins, and in automobiles. Face masks were more likely to be thrown away in urban, sub-urban, and rural areas, according to the current study [7]. Toxic concentrations of biowaste and medical waste fuel led sudden rise in plastic pollution by the COVID-19 pandemic. Reusable masks are more breathable and porous than surgical masks or respirators, yet they are less effective in dangerous conditions. Only 5% of the single-use face masks given during the COVID-19 epidemic were reprocessed using multi-use masks, according to a study. Face mask pretreatment, which entails separating plastic materials and disinfecting them, is a time-consuming and expensive operation. In view of the largenumber of face masks and PPE kits created during the COVID-19 epidemic, it is vital that more attention be given to developing a circular economy [8].

Salivation droplets are a primary mode of transmission for Coronaviruses. As a result, containing the virus at an early stage would be a large win. The use of a face mask can help prevent the spread of the disease both within and outside of the mask. It is helpful to wear face masks to keep the illness from spreading [9].Masks are the best form of protection in public places and anywhere else. All of these measures were put in place as precautionary measures to prevent the spread of the disease, which included social distancing, mandatory face masks in homes, quarantine, restrictions on citizens' travel, and the cancellation or exclusion of large social events and meetings. As a precaution, people with elevated body temperatures should not be permitted into public spaces and should wear a mask while they are there in order to prevent infection and spread of the virus.

Temperature and mask inspections must be conducted at all city gateways, offices, shopping malls, and hospital entrances. In the wake of this research, a door-opening device that automatically detects human body temperature was developed. Mask detection, the number of individuals present, and the temperature are all integrated into this system's revolutionary concept [10].

LITERATURE SURVEY

Das *et al.* claimed that nodoubt Covid-19 has had an enormous influence on the economics, psyche, and well-being of the human species as a whole. COVID-19 can be prevented by wearing a face mask in public. The form and categorization of masks are explored in this study. There is a global shortage of N95 and medical masks. To be safe against the coronavirus epidemic, it is recommended that everyone uses a mask whenever feasible [11]. By blocking transmission from an infected individual or protecting healthy persons who wear the mask, a face mask may help prevent the spread of a virus [12]. Face masks, particularly FFPs worn by medical staff, are in high demand during the COVID-19 epidemic. The effectiveness of single and combination NPIs in preventing the transmission of COVID-19 was evaluated using relative risk assessment and statistical modeling, with an emphasis on various face mask designs and operating features. FFP3 respirators that are not disposable are not included in this review [13].

Moreno *et al.*, in the context of COVID-19 discussed that face mask use significantly contributes to global plastic trash and MP inhalation, and this needs to be addressed. As facemasks have grown more prevalent, research on the health effects of prolonged exposure to particulate matter (MP) is warranted [14]. In a study by Chen *et al.*, it was found that polypropylene nonwoven fabric fibers were responsible for the majority of microplastics released from the face mask. Abrasion and aging from the usage of face masks have increased the release of microplastics due to an increase in medium and blue microplastics in recent years. Face masks may accumulate microplastics from the air while they are in use. In addition to helping with pandemic control, properly disposing of unused masks is also good for environmental safety and pandemic preparedness. The effects of various new and used DFMs on ecosystems' microplastics release should be evaluated before using any of these DFMs [15]. Colloid and interface science can be used to explain how face masks work to protect against airborne diseases. Look at the current design and production requirements and recommendations for face masks for both the medical profession and for the general population [16]. According to Morgana *et al.*, plastic water contamination from single-use face masks during and after the Ebola pandemic is a major environmental concern. A face mask material's ability to release plastic is determined by the amount of mechanical shear stress applied to the material [17].

Xie *et al.*, identified 12 phthalatesin 56 mask samples from around the world. Because they minimize shear and increase adaptability, phthalates are frequently used in polymer manufacture as plasticizers. Phthalateexposures for adults and children were compared by calculating phthalate intake from masks on a daily basis. Finally, the dangers of phthalate exposure in terms of cancer and non-cancerous consequences were assessed. In the fight against the epidemic, face masks have made a substantial contribution, and masks will continue to be worn for a long time to come [18]. Daniels *et al.*, studied exhaled breath condensate (EBC), which can be conveniently collected utilizing mask-based sample equipment, and an electrochemical biosensor with a modular architecture, can be used to detect and quantify COVID-19. SARS-CoV-2 was found in five of the seven people who were tested positive for the virus. With this, it will be possible to do a rapid and targeted examination of the infectious status, potentially even at home [19].

Gorkem Tutal Gursoy *et al.*, looked into how COVID-19 masks and disinfectants were affecting migraine sufferers. More than 300 migraine sufferers participated in the research. A face-to-face survey was utilized to obtain demographic information, migraine characteristics, and the usage of masks and disinfectants. Because each person's social and vocational situations are unique, it is important for migraine patients to be educated about the best ways to prevent migraines. A year after the first year of the COVID-19 pandemic, look at how migraine symptoms have changed and see if different types of masks or disinfectants or the frequency of use has an effect on this. Almost all of our patients didn't keep a migraine diary, and the study population was disproportionately female. Migraine sufferers have long-term pandemic obligations. Using a mask for an extended period of time has been associated with an increase in the frequency and intensity of migraine attacks. An appropriate protective method tailored to the needs of migraine sufferers can improve both their quality of life and cost-effectiveness of the treatment [20].

Kumar *et al.* have come out with face mask categorization and detection technologies in the previous two years. Methods for upgrading the present technology and developing face mask detectors in distant and obstructed images are generally lacking in the literature. We know that a wide variety of object detection subfields can benefit from using the YOLO algorithm series. Face mask identification is difficult for these algorithms when the mask is veiled and distant, or when the mask is too tiny to detect. This paper presents ETL-YOLO v4 as an upgraded version of the current small YOLO v4 technique that combines extra convolutional and detection layers to increase face mask detection accuracy. The ETL-YOLO v4 can do real-time detection with the use of a ZED or CCTV camera. Monitoring whether or not individuals are using face masks in public areas, hospitals, schools, workplaces, and universities during the current COVID-19 epidemic may be done using the ETL-YOLO v4 by governments and healthcare institutions. We offer a new approach that starts with synthetic face mask samples and then feeds them into a generative adversarial network for training with multiple points of view *i.e.* the Generation of adversarial networks [21]. In high-traffic places like airports and supermarkets, Zeng *et al.* claim that intelligent mask monitoring can be carried out by deploying the necessary equipment, which can considerably boost COVID-19 pandemic prevention and control efficiency. The FMD-Yolo architecture and algorithm are designed to recognize and detect facemasks automatically. An approach for detecting face masks, called FMD-Yolo, has been suggested, which includes feature extraction, feature fusion, and post-processing. It is possible to reliably identify people who have worn a mask using FMD-Yolo, a real-world application [22].

Batliner *et al.*, conducted an analysis of the Interspeech computational challenge series' Mask Sub-Challenge (or MSC) and the results were published. It was the goal of this challenge to come up with ways to tell if a speaker was wearing a surgical mask based only on audio. " Whether people have worn surgical masks or not can be recorded in the MASC database that is used by the people speaking both structured and unstructured text. When people wear masks, programs can only rely on a limited set of voice biometrics, according to the research [23]. Fernandez-Arribas *et al.*, suggested a general rule of thumb, stating that FFP2 masks should be used in poorly ventilated indoor public settings where virus inhalation is a concern. OPE was analysed in various types of masks to discover how long and how frequently they are used to see how they affect human health [24].

According to Kirchbuchner *et al.*, COVID-19 can be prevented by using face masks. Facial recognition (FR) is made more difficult by masks because they hide crucial distinguishing qualities. Facial recognition systems are more vulnerable to reveal assaults than disguised attacks. This means that FR systems are more vulnerable to veiled and unsuspecting assaults when the threshold is determined when unmasked comparisons are made. Attacks on FR systems with masked references are harder to conduct than attacks on systems without them. False faces wearing masks (AM2) pose a greater threat to FR systems than an actual assault with a masked face (AM1). Because of this, there are questions about the safety of FR systems in the event of sabotage [25].

Loey *et al.*, used deep and traditional machine learning to identify face masks. There are two parts of the model that we've come up with. Initially, the system is designed to retrieve Resnet50 characteristics. Second, component face masks are classified using decision trees, support vector machines (SVMs), and an ensemble method. To conduct this study, researchers gathered data from three separate sets of subjects who had been masking their faces. These are three datasets: the Real-World Masked Face (RMFD), the Simulated Masked Face (SMFD), and the Labeled Faces in the Wild (LFW) (LFW). The RMFD's SVM classifier scored a perfect score of 99.64 percent throughout testing. According to SMFD, it was 99.49% correct, while according to LFW, it was 100% accurate. There is a trade-off between efficiency and accuracy in the vast majority of traditional machine-learning methods [26]. Vaquer *et al.* for the first time, developed a noninvasive approach to diagnose COVID-19. In order to detect SARS-CoV-2 antigens trapped in surgical face masks, nanoparticle transfer biosensors are utilized even when tested on asymptomatic persons and masks worn for 30 minutes, the biosensors had outstanding sensitivity and specificity. Samples may be collected using a face mask for 30 minutes instead of the more time-consuming nose swab method, and the findings can be sent in as little as ten minutes. The biosensor signal weakens if the diagnosis is delayed, increasing the risk of a negative outcome [27]. According to Yalcin *et al.*, these face masks, which have been widely worn since the emergence of COVID-19 and damage environment, were sliced into little pieces and mixed with bitumen binder. One ratio of SBS (styrene–butadiene styrene) was mixed with five different ratios of the waste mask, and the physical, chemical, and rheological properties of the modified binder were examined. Increased binder softening point and viscosity, as well as lower penetration values were observed, resulting from the addition of waste mask and SBS to pure bitumen [28]. Xue *et al.*, proposed an intelligent face mask capable of detecting coronavirus aerosols. To detect coronavirus particle aerosols, the face mask incorporates a nanowire sensor for aerosol-mediated diagnostics. In addition, it does not require the use of skilled medical workers, generates no potentially infectious waste, and may be performed nearly anywhere and at any time. It is particularly beneficial in situations when speedy screening is needed, such as customs and airports, where a large number of suspected cases can overwhelm medical resources [29].

A research by Sethi *et al.* focused on the development of a real-time and highly accurate method for detecting people who aren't wearing masks in public. Low inference time and high accuracy are the goals of the new method, which consists of a combination of one- and two-stage detectors. An initial base for many feature maps was the ResNet50 neural network, which was used as a starting point for transfer learning. For better localization results, we recommend using a bounding box transformation for mask detection. As a starting point, the experiment employed three widely used models: ResNet50, AlexNet, and MobileNet. The use of ResNet50 is a recommended way to achieve a high degree of precision (98.2 percent). Mask detection accuracy and recall of this new model surpass the Retina Face Mask Detector public baseline model by 11.07% and 6.44 percent, respectively. The excellent performance of the suggested type makes it perfect for video surveillance systems [30]. A study by Loey *et al.*, aimed at annotating and locating medical face mask objects using real-world images. COVID-19 transmission is prevented in public locations where people use medical face masks. The proposed model is divided into two components. For feature extraction, the ResNet-50 deep learning model serves as the foundation of the first component. On top of YOLO v2, a second component is needed to distinguish medical face masks. When the anchor boxes are estimated using a meanIoU, the detection performance is improved. Two optimizers are used during training to achieve the best results possible. Input photo of a face covered by a mask and the ResNet-50-based YOLO-v2 model can identify it [31]. Fashoto *et al.*, proposed deep learning systems like Inceptionv3 that were able to detect Covid-19 masks with 99.9% accuracy. There is a widespread problem with the datasets used to detect face masks since they are wrongly generated and do not accurately represent real-world settings. To replicate deep learning systems, COVID-19 face mask pictures must be utilized. A new study finds that larger training parameters in deep learning architectures like YOLO, Xception, and DenseNet have yet to boost face mask identification [32]. Prusty *et al.*, claimed that mask detection is a great paradigm for detecting masks when they are needed. In the current epidemic of COVID-19, people who use masks in public must be properly monitored. This automated technique would prevent the transmission of the sickness to security officers or others who are assigned to mask monitoring. This study proposes a novel method of masking recognition that might be employed in an automated system. Data enhancement by filtering improves model performance in comparison to traditional MDM techniques. The YOLOv3 model, a deep learning object identification tool, is utilized to generate a mask detection model in order to enrich an existing conventional dataset. The data is being supplemented using image filtering techniques like grayscale and Gaussian blur. The mask detection object detection model may be trained using this larger dataset. [more] Based on the data enhancement, a 99.8 percent accuracy rate for the mask detection model was found. In the last stage, we compare the results of the traditional and augmented data approaches that we've come up with. With the conventional mask detection model, we found an average degree of confidence of 0.94 when applied to images and videos of individuals (type A), groups (type B), and a video of a group (type C). The mask detection model based on data augmentation has an average confidence level of 0.97, 0.96, and 0.93, respectively, for types A, B, and C. In regard to using mask detection models, the suggested data augmentation-based approach is preferable to the usual mask detection model [33] in this work. Tiwari *et al.*, proposed machine learning techniques such as TensorFlow, Keras, OpenCV, and Scikit Learn that can help achieve this aim. The suggested method is able to identify the face even if it is obscured by a mask in an image or video. This device is capable of recognizing a person's face even while they are wearing a mask. This is a very labor-intensive process [34]. Decker *et al.*, claimed that improved detection and quantification of SARS-CoV-2 are now possible. Higher FMS viral levels are connected to more severe illness than higher NPS viral loads. Similar to NPS, the viral load of FMS was the greatest early in the disease and in people with active respiratory symptoms, indicating that FMS may have a role in the transmission of the virus [35].

The paper's primary goal will be to warn the authorities so that COVID-19 can be contained if someone is caught without a face mask. CCTV cameras recorded the image used in the procedure. A neural network is used to extract features and classifications from the data (NN). An accuracy of 99.0 percent has been achieved by the model that has been trained. The suggested technique uses the MobileNetV2 pre-trained classifier for classification. Photos of actual individuals are used in a Kaggle dataset. The approach included training, validating, and testing the model. In order to better understand the structure of the paper, below is the breakdown. There is an introduction to the literature in Section 2. According to the study's third portion, the methodology is explained. Both Sections 4 and 5 provide an explanation of the suggested method. Finally, in Section 6, the suggested model is described.

RESEARCH METHODOLOGIES

The CNN, a Deep Learning algorithm, reinterprets each portion of the picture in the image. The most common use of CNNs in deep learning is to evaluate visual information. For picture recognition and classification, CNNs are employed. All neurons in the CNN are connected to one another and to the storage layer, which is made up of all neurons in the funnel-shaped network. Traditional classification methods are increasingly being replaced by CNNs in order to better collect picture information and improve classification performance.

Dataset

The study's dataset was stored in JPEG, PNG, and other picture file formats. Data and photographs from open-source projects, including Kaggle's face mask detection dataset, were available. The outcome was a wide variety of image sizes and resolutions to select from. Finally, the data was divided into two categories: "With" and "Without" the mask. There is a GitHub repository for data collecting.

PROPOSED METHODOLOGY

Conv Net is the model's foundation for determining whether or not someone is wearing a mask. One of the four phases represented in Fig. (**2**) is face mask detection.

Data Collection

Three different datasets combined 1440 photographs for mask detection. Another 150 photographs were taken using smartphones, webcams, and CCTV cameras. A combination of RGB-formatted CCTV and webcam footage was used to identify video masks. Training and testing our model on the Real-World Masked Face Dataset and the Simulated Masked Face Dataset helped us to avoid overfitting the models (SMFD).

Preprocessing and Augmentation of Data

Because photographs in the dataset were not of the same size, preprocessing was required for this experiment. The training of deep learning models necessitates a large amount of data. Data cleansing is performed to eliminate any erroneous photos from the collection. All photos were resized to 256 × 256 pixels using the Keras Image Data Generator technique. This was the final step before we could normalize all photos. After then, the photos are classified as having or not having masks. A NumPy array is created from the image array for the ease of computation. In addition, MobileNetV2's preprocess input function is also utilized. After that, the training dataset's quality and quantity are both improved using the data augmentation technique.

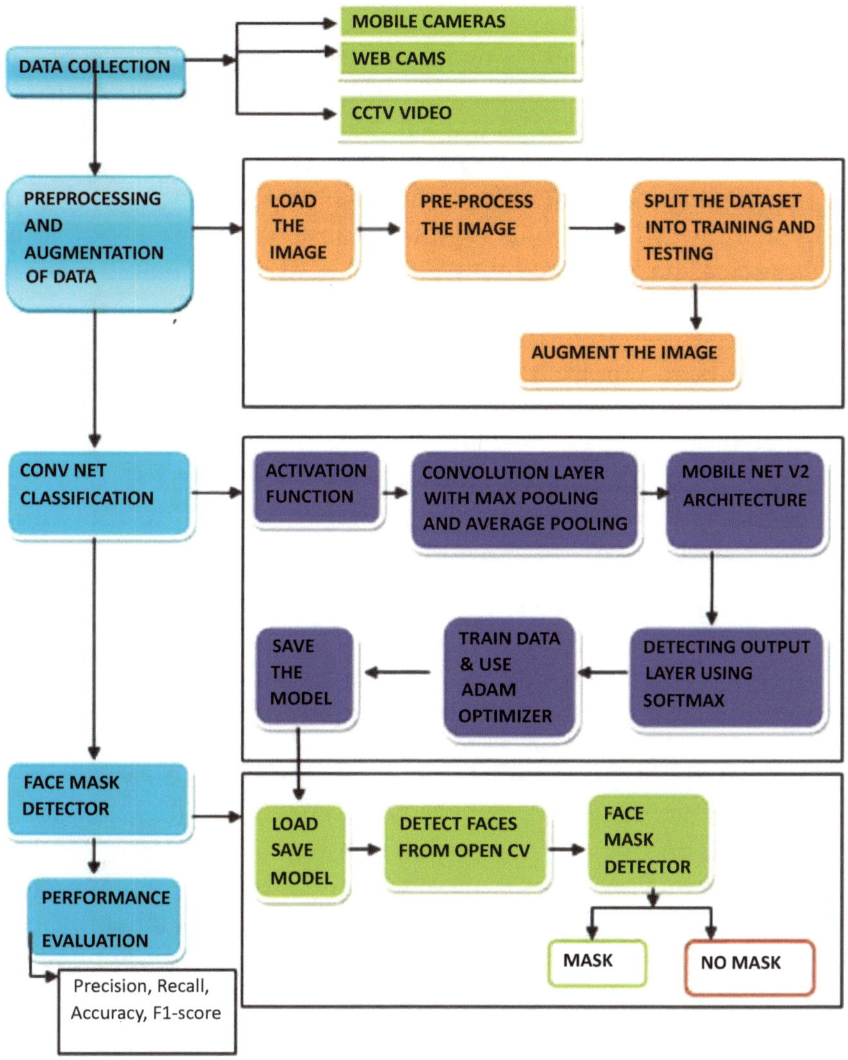

Fig. (2). Architecture of face mask detection.

Proposed Conv Net Architecture

Conv Net is employed in the classification and processing of images. One or more convolutional layers make up a CNN. CNN prefers to focus on a certain area of a photograph rather than the complete scene. CNN has both an input and an output layer in its construction. Activation function, convolutional layer, pooling layer, and Mobile v2 are all part of this architecture. The activation functions of a neuron determine whether or not it fires. The quality of model learning is greatly determined by the hidden and output layers' choice of the activation function. The Active layers are often activated using the ReLU activation algorithm. A smaller feature map results from combining many layers. Reduced parameter training speeds up the computations without compromising the system's quality while still allowing for more parameters to be trained. Maximal and average pooling are the two most prevalent types of pooling processes. Max pooling is the technique of saving the kernel's most critical value . It is also important to note that pooling is based on averaging all data in a region. MobileNetV2 is a powerful picture classification tool. Data is entered into Softmax and the probability distribution is generated for the whole set of values entered. Adam's stochastic gradient descent approach enhanced the training accuracy. The model can recognize masks in any image after it has been taught. To begin, each person in the picture is identified using a face detection model. Finally, the model for detecting masks of faces is fed, this data is shown in Fig. (**2**). If the model found any patterns or traits in a photo, it would be labeled "Mask" or "No Mask." Using the f1-score, precision, recall, and accuracy scores, we compared the two models' results after training and testing.

Four Convolution Layer with Max and Average Pooling

There are two types of pooling methods: average pooling and max pooling. Keras is the starting point for our Conv Net model (). The Max pooling and average pooling processes are employed before the Relu activation feature in the first hidden layer as shown in Fig. (**3**). With the use of max pooling, we can collect a lot of data while also shrinking the photos. The data is then passed on to the subsequent convolution layers, which include the second, third, and fourth. In order to obtain the most noteworthy data, we go back to using maximum pooling. Then, it's flattened and trained on the photo matrix as well. Flattening and training the photo matrix are as follows. Instead of using the max pooling function, we employed the average pooling process to see how the model performed. For more precise training, we used Adam stochastic gradient descent technique. Our dataset containing 85 percent of the images we use for training is shown in Fig. (**3**).

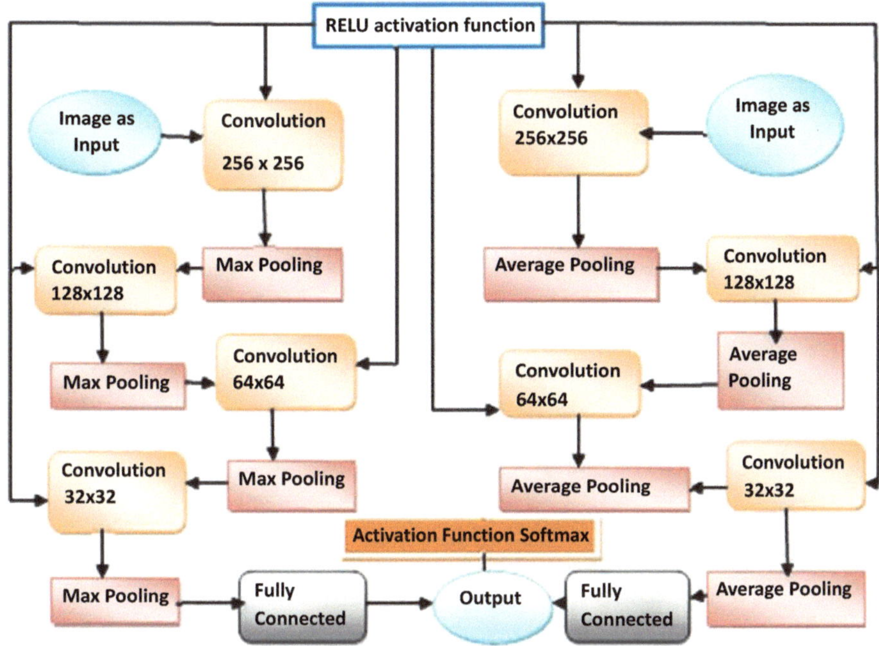

Fig. (3). Four Convolution Layer with Max and Average Pooling.

MobileNetV2 Architecture

Deep learning model MobileNetV2 is a lightweight CNN that leverages TensorFlow's picture weights to train. The model examines our photos and extracts the most salient details. OpenCV-based ResNet-10 was used to build the initial model. With OpenCV, faces and masks may be detected and analyzed in photos and video feeds. A typical pooling operation uses a 32-layer pool as its standard size. There are two types of activation functions: Relu and Softmax. Precision was improved by using a learning rate of 0.01.

Face Mask Detector Face Detection Using Open-CV

Using TensorFlow and Keras, we were able to create this model. For training, 85 percent of the dataset is used, while the remaining 15 percent is used for testing. Pre-processing has been used to enhance the visual attractiveness of this photograph. It is necessary to use Flatten and Dense to link up the various sections of your image. The face detection model is pretrained on 400x400 photos with 150,000 iterations. Due to the fact that this detector uses only one shot, it is more accurate and faster than the Dual Shot Detector. Both static photos and real-time video feeds can benefit from our face recognition model. In order to recognize masks in images, the model has been trained and is ready for usage. Using the face detection methodology, all people in an image are first identified. After this, a CNN-based face mask detection technique is applied. Mask or No Mask? Images would be categorized based on the model's ability to extract hidden patterns and features from them.

Using a Performance Matrix to Evaluate Performance

Evaluations based on precision, recall, F1 scores, and accuracy were carried out after training and testing. The following formulae were employed:

$$Precision = \frac{TP}{TP + FP}$$
$$Recall = \frac{TP}{TP + FN}$$
$$Accuracy = \frac{TP + TN}{TP + FP + TN + FN}$$

$$F1 - score = \frac{Recall * Precision}{Recall + Precision}$$

RESULTS AND ANALYSIS

Co-authors gathered 130 photographs for the study, which contained 1940 images from various sources, using a webcam and a cell phone camera. Table **1** displays the feature map's accuracy when the Deep CNN model with Max Pooling and Average Pooling is used. Max Pooling can only provide a 96% training success rate and a 98% testing success rate. It is possible to reduce the size of the feature map by using average pooling. The anticipated results are thus less accurate than those obtained *via* maximum pooling. Based on the anticipated results, Table **1** displays a training accuracy of 93% and a validation accuracy of 95%.

Table 1. Outcomes of Max Pooling and Average Pooling.

Epoch	Max Pooling of Different Epochs (%)				Average Pooling Different Epochs (%)			
	Training Loss	Training Accuracy	Validation Loss	Validation Accuracy	Training Loss	Training Accuracy	Validation Loss	Validation Accuracy
1	43.14	90.77	13.33	97.74	44.55	89.93	14.33	90.13
2	11.02	92.88	9.44	95.35	12.81	91.22	10.44	91.13
3	9.46	94.98	5.34	97.11	11.90	91.86	10.34	92.02
4	9.22	95.54	5.26	97.26	10.83	92.07	9.26	92.53
5	8.05	95.99	7.11	98.04	10.22	92.25	9.11	94.26
6	7.9	96.13	6.36	98.24	9.96	93.38	9.36	95.22
7	7.84	96.25	6.13	98.55	9.72	93.70	8.13	95.40

Improved accuracy was achieved by employing the MobileNetV2 architecture. Table **2** illustrates the validation and test accuracy for every period.

Table 2. Outcomes of MobileNetV2 .

Epoch	Training Loss	Training Accuracy	Validation Loss	Validation Accuracy
1	5.44%	99.68%	5.22%	99.72%
2	5.33%	99.73%	5.13%	99.82%
3	5.22%	99.82%	5.10%	99.33%
4	5.13%	99.25%	4.90%	99.11%
5	4.91%	99.02%	4.73%	99.14%
6	4.83%	99.14%	4.62%	99.26%
7	4.79%	99.85%	4.57%	99.90%

Table **2** shows that Mobile Net V2 has a precision of 99.85 percent for training data and a validity precision of 99.90 percent. MobilenetV2 consistently distinguishes between "Mask" and "No Mask" video broadcast as shown in Fig. (**4**).

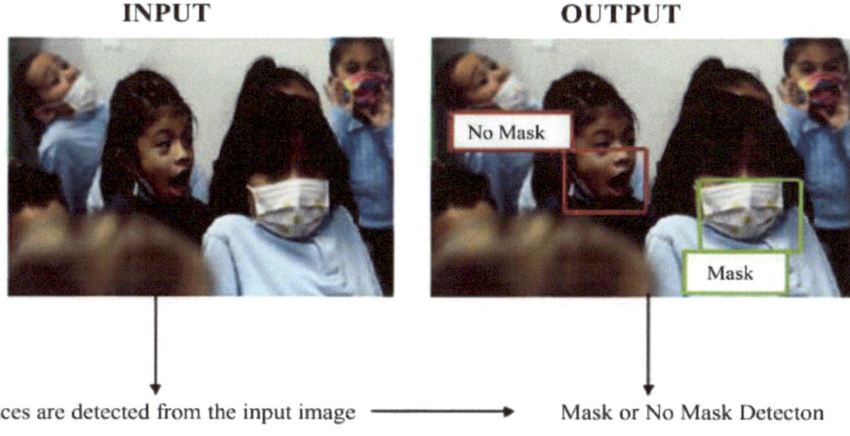

Fig. (4). Mask or No Mask detection model in crowded places using MobileNetV2.

CONCLUSION

This study describes a face mask identification system that uses both static images and live video to potentially stop the COVID-19 infection from spreading. The usage of a face mask is determined using Keras, OpenCV, and Conv Net. Our main goal was to develop an accurate mask identification model. This may be done fast and accurately by using a face mask recognition system. Mask-wearers will be separated from the rest of the group.

REFERENCES

[1]   A.K. Sharadhi, V. Gururaj, S.P. Shankar, M.S. Supriya, and N.S. Chogule, "Face mask recogniser using image processing and computer vision approach", *Glob. Transit. Proc.,* vol. 3, no. 1, pp. 67-73, 2022.
[http://dx.doi.org/10.1016/j.gltp.2022.04.016]

[2]   Puja Gupta, Varsha Sharma, and Sunita Varma, "A novel algorithm for mask detection and recognizing actions of human", *Expert Systems with Applications,* p. 116823, 2022.

[3]   Mary Abboah-Offei, Yakubu Salifu, Bisi Adewale, Jonathan Bayuo, Rasheed Ofosu-Poku, and Edwina Beryl AddoOpare-Lokko, "A rapid review of the use of face mask in preventing the spread of COVID-19", *International journal of nursing studies advances,* vol. 3, p. 100013, 2021.

[4]   Anna Zochowska, Pawel Jakuszyk, Maria M. Nowicka, and Anna Nowicka, "Are covered faces eye-catching for us? The impact of masks on attentional processing of self and other faces during the COVID-19 pandemic", *cortex,* vol. 149, pp. 173-187, 2022.

[5]   D.D. Bussan, L. Snaychuk, G. Bartzas, and C. Douvris, "Quantification of trace elements in surgical and KN95 face masks widely used during the SARS-COVID-19 pandemic", *Sci. Total Environ.,* vol. 814, p. 151924, 2022.
[http://dx.doi.org/10.1016/j.scitotenv.2021.151924] [PMID: 34838548]

[6]   Y.T. Tesfaldet, and N.T. Ndeh, "Assessing face masks in the environment by means of the DPSIR framework", *Sci. Total Environ.,* vol. 814, p. 152859, 2022.
[http://dx.doi.org/10.1016/j.scitotenv.2021.152859] [PMID: 34995587]

[7]   E.E.Y. Amuah, E.P. Agyemang, P. Dankwa, B. Fei-Baffoe, R.W. Kazapoe, and N.B. Douti, "Are used face masks handled as infectious waste? Novel pollution driven by the COVID-19 pandemic", *Resources, Conservation & Recycling Advances,* vol. 13, p. 200062, 2022.
[http://dx.doi.org/10.1016/j.rcradv.2021.200062]

[8]   S.S. Ray, H. Lee, D. ThiThanhHuyen, S. Chen, and Y. Kwon, "Microplastics waste in environment: A perspective on recycling issues from PPE kits and face masks during the COVID-19 pandemic", In: *Environmental Technology & Innovation,* 2022, p. 102290.

[9]   M.A. Chowdhury, M.B.A. Shuvho, M.A. Shahid, A.K.M.M. Haque, M.A. Kashem, S.S. Lam, H.C. Ong, M.A. Uddin, and M. Mofijur, "Prospect of biobased antiviral face mask to limit the coronavirus outbreak", *Environ. Res.,* vol. 192, p. 110294, 2021.
[http://dx.doi.org/10.1016/j.envres.2020.110294] [PMID: 33022215]

[10] B. Varshini, H.R. Yogesh, S.D. Pasha, M. Suhail, V. Madhumitha, and A. Sasi, "IoT-Enabled smart doors for monitoring body temperature and face mask detection", *Glob. Transit. Proc.,* vol. 2, no. 2, pp. 246-254, 2021.
[http://dx.doi.org/10.1016/j.gltp.2021.08.071]

[11] S. Das, S. Sarkar, A. Das, S. Das, P. Chakraborty, and J. Sarkar, "A comprehensive review of various categories of face masks resistant to Covid-19", *Clin. Epidemiol. Glob. Health,* vol. 12, p. 100835, 2021.
[http://dx.doi.org/10.1016/j.cegh.2021.100835] [PMID: 34368502]

[12] C. Raina MacIntyre, V. Costantino, and A. Chanmugam, "The use of face masks during vaccine roll-out in New YorkCity and impact on epidemic control", *Vaccine,* vol. 39, no. 42, pp. 6296-6301, 2021.
[http://dx.doi.org/10.1016/j.vaccine.2021.08.102] [PMID: 34538699]

[13]   N.J. Rowan, and R.A. Moral, "Disposable face masks and reusable face coverings as non-pharmaceutical interventions (NPIs) to prevent transmission of SARS-CoV-2 variants that cause coronavirus disease (COVID-19): Role of new sustainable NPI design innovations and predictive mathematical modelling", *Sci. Total Environ.*, vol. 772, p. 145530, 2021.
[http://dx.doi.org/10.1016/j.scitotenv.2021.145530] [PMID: 33581526]

[14]   A. Torres-Agullo, A. Karanasiou, T. Moreno, and S. Lacorte, "Overview on the occurrence of microplastics in air and implications from the use of face masks during the COVID-19 pandemic", *Sci. Total Environ.*, vol. 800, p. 149555, 2021.
[http://dx.doi.org/10.1016/j.scitotenv.2021.149555] [PMID: 34426330]

[15]   X. Chen, X. Chen, Q. Liu, Q. Zhao, X. Xiong, and C. Wu, "Used disposable face masks are significant sources of microplastics to environment", *Environ. Pollut.*, vol. 285, p. 117485, 2021.
[http://dx.doi.org/10.1016/j.envpol.2021.117485] [PMID: 34087638]

[16]   M. Liao, H. Liu, X. Wang, X. Hu, Y. Huang, X. Liu, K. Brenan, J. Mecha, M. Nirmalan, and JianRen. Lu., "A technical review of face masks wearing in preventing respiratory COVID-19 transmission", *Curr. Opin. Colloid Interface Sci.*, vol. 52, p. 101417, 2021.
[http://dx.doi.org/10.1016/j.cocis.2021.101417] [PMID: 33642918]

[17]   S. Morgana, B. Casentini, and S. Amalfitano, "Uncovering the release of micro/nanoplastics from disposable face masks at times of COVID-19", *J. Hazard. Mater.*, vol. 419, p. 126507, 2021.
[http://dx.doi.org/10.1016/j.jhazmat.2021.126507] [PMID: 34323718]

[18]   H. Xie, W. Han, Q. Xie, T. Xu, M. Zhu, and J. Chen, "Face mask—A potential source of phthalate exposure for human", *J. Hazard. Mater.*, vol. 422, p. 126848, 2022.
[http://dx.doi.org/10.1016/j.jhazmat.2021.126848] [PMID: 34403943]

[19]   J. Daniels, S. Wadekar, K. DeCubellis, G.W. Jackson, A.S. Chiu, Q. Pagneux, H. Saada, I. Engelmann, J. Ogiez, D. Loze-Warot, R. Boukherroub, and S. Szunerits, "A mask-based diagnostic platform for point-of-care screening of Covid-19", *Biosens. Bioelectron.*, vol. 192, p. 113486, 2021.
[http://dx.doi.org/10.1016/j.bios.2021.113486] [PMID: 34260968]

[20]   H. Yuksel, S.G. Kenar, G.T. Gursoy, and H. Bektas, "The impacts of masks and disinfectants on migraine patients in the COVID-19 pandemic", *J. Clin. Neurosci.*, vol. 97, pp. 87-92, 2022.
[http://dx.doi.org/10.1016/j.jocn.2022.01.006]

[21]   A. Kumar, A. Kalia, and A. Kalia, "ETL-YOLO v4: A face mask detection algorithm in era of COVID-19 pandemic", *Optik,* vol. 259, p. 169051, 2022.
[http://dx.doi.org/10.1016/j.ijleo.2022.169051] [PMID: 35411120]

[22]   P. Wu, H. Li, N. Zeng, and F. Li, "FMD-Yolo: An efficient face mask detection method for COVID-19 prevention and control in public", *Image Vis. Comput.*, vol. 117, p. 104341, 2022.
[http://dx.doi.org/10.1016/j.imavis.2021.104341] [PMID: 34848910]

[23]   M.M. Mohamed, M.A. Nessiem, A. Batliner, C. Bergler, S. Hantke, M. Schmitt, A. Baird, A. Mallol-Ragolta, V. Karas, S. Amiriparian, and B.W. Schuller, "Face mask recognition from audio: The MASC database and an overview on the mask challenge", *Pattern Recognit.*, vol. 122, p. 108361, 2022.
[http://dx.doi.org/10.1016/j.patcog.2021.108361] [PMID: 34629550]

[24]   J. Fernández-Arribas, T. Moreno, R. Bartrolí, and E. Eljarrat, "COVID-19 face masks: A new source of human and environmental exposure to organophosphate esters", *Environ. Int.*, vol. 154, p. 106654, 2021.
[http://dx.doi.org/10.1016/j.envint.2021.106654] [PMID: 34051653]

[25]   M. Fang, N. Damer, F. Kirchbuchner, and A. Kuijper, "Real masks and spoof faces: On the masked face presentation attack detection", *Pattern Recognit.*, vol. 123, p. 108398, 2022.
[http://dx.doi.org/10.1016/j.patcog.2021.108398] [PMID: 34720199]

[26]   M. Loey, G. Manogaran, M.H.N. Taha, and N.E.M. Khalifa, "A hybrid deep transfer learning model

with machine learning methods for face mask detection in the era of the COVID-19 pandemic", *Measurement,* vol. 167, p. 108288, 2021.
[http://dx.doi.org/10.1016/j.measurement.2020.108288] [PMID: 32834324]

[27]   A. Vaquer, A. Alba-Patiño, C. Adrover-Jaume, S.M. Russell, M. Aranda, M. Borges, J. Mena, A. del Castillo, A. Socias, L. Martín, M.M. Arellano, M. Agudo, M. Gonzalez-Freire, M. Besalduch, A. Clemente, E. Barón, and R. de la Rica, "Nanoparticle transfer biosensors for the non-invasive detection of SARS-CoV-2 antigens trapped in surgical face masks", *Sens. Actuators B Chem.,* vol. 345, p. 130347, 2021.
[http://dx.doi.org/10.1016/j.snb.2021.130347] [PMID: 34188360]

[28]   E. Yalcin, A. Munir Ozdemir, B. Vural Kok, M. Yilmaz, and B. Yilmaz, "Influence of pandemic waste face mask on rheological, physical and chemical properties of bitumen", *Constr. Build. Mater.,* vol. 337, p. 127576, 2022.
[http://dx.doi.org/10.1016/j.conbuildmat.2022.127576] [PMID: 35493028]

[29]   Q. Xue, X. Kan, Z. Pan, Z. Li, W. Pan, F. Zhou, and X. Duan, "An intelligent face mask integrated with high density conductive nanowire array for directly exhaled coronavirus aerosols screening", *Biosens. Bioelectron.,* vol. 186, p. 113286, 2021.
[http://dx.doi.org/10.1016/j.bios.2021.113286] [PMID: 33990035]

[30]   S. Sethi, M. Kathuria, and T. Kaushik, "Face masks detection using deep learning: An approach to reduce risk of Coronavirus spread", *J. Biomed. Inform.,* vol. 120, p. 103848, 2021.
[http://dx.doi.org/10.1016/j.jbi.2021.103848] [PMID: 34171485]

[31]   M. Loey, G. Manogaran, M.H.N. Taha, and N.E.M. Khalifa, "Fighting against COVID-19: A novel deep learning model based on YOLO-v2 with ResNet-50 for medical face mask detection", *Sustain Cities Soc.,* vol. 65, p. 102600, 2021.
[http://dx.doi.org/10.1016/j.scs.2020.102600] [PMID: 33200063]

[32]   E. Mbunge, S. Simelane, S.G. Fashoto, B. Akinnuwesi, and A.S. Metfula, "Application of deep learning and machine learning models to detect COVID-19 face masks - A review", *Sustainable Operations and Computers,* vol. 2, pp. 235-245, 2021.
[http://dx.doi.org/10.1016/j.susoc.2021.08.001]

[33]   M.R. Prusty, V. Tripathi, and A. Dubey, "A novel data augmentation approach for mask detection using deep transfer learning", *Intelligence-Based Medicine,* vol. 5, p. 100037, 2021.
[http://dx.doi.org/10.1016/j.ibmed.2021.100037]

[34]   G. Kaur, R. Sinha, P.K. Tiwari, S.K. Yadav, P. Pandey, R. Raj, A. Vashisth, and M. Rakhra, "Face mask recognition system using CNN model", *Neuroscience Informatics,* p. 100035, 2021.

[35]   C.M. Williams, D. Pan, J. Decker, A. Wisniewska, E. Fletcher, S. Sze, S. Assadi, R. Haigh, M. Abdulwhhab, P. Bird, C.W. Holmes, A. Al-Taie, B. Saleem, J. Pan, N.J. Garton, M. Pareek, and M.R. Barer, "Exhaled SARS-CoV-2 quantified by face-mask sampling in hospitalised patients with COVID-19", *J. Infect.,* vol. 82, no. 6, pp. 253-259, 2021.
[http://dx.doi.org/10.1016/j.jinf.2021.03.018] [PMID: 33774019]

CHAPTER 5

Wireless Ph Sensor Employing Zigbee 3.0 Protocol

Jacob Abraham[1] and S. Kannadhasan[2,*]

[1] *Department of Electronics, B P C College, Piravom P.O 686664, Kerala, India*

[2] *Department of Electronics and Communication Engineering, Study World College of Engineering, Coimbatore, Tamil Nadu-641105, India*

Abstract: PH plays an important role in determining product quality in industries like various chemical, petrochemical, petroleum refineries, fertilizer, pharmaceutical, food industries, effluent treatment, and in many other organic and inorganic plants. For instance, in any industrial wastewater treatment plant, PH is monitored and controlled by manipulating the acid or base stream which is a strong acid or strong base. Modern treatment plant involves physical and chemical precipitation/flocculation along with biological treatment in aerators/trickle filters, membranes, *etc*, where the control of PH is the key factor for efficient treatment. In chemistry, PH is ameasure of the acidity or basicity of an aqueous solution. Pure water is said to be neutral, with a PH close to 7.0 at 25 degree Celsius. Solutions with a PH less than 7 are said to be acidic and solutions with a PH greater than 7 are basic or alkaline. PHmeasurements are important in medicine, biology, chemistry, agriculture, forestry, *etc*. By PH control we mean to maintain the PH value during continuous operation at a specific desired value through manipulating the alkaline flow rate. Usually in most industrial applications, the desired value is chosen to be around 7. This is the safest value for portable water, utility water used in industry, or waste disposing water.

Keywords: Agriculture, Filter, PH, Sensors, Zigbee.

INTRODUCTION

The working of this PH controlling system is as follows: firstly the pH electrode senses the PH value of the solution that transfers it to the AVR. In the microcontroller, we can set a desired set point value. If the output of the pH electrode is less than the set point value, and the behaviour of the solution is acidic, then compare the PH value of the solution with the set point value. Then the microcontroller transmits an electrical signal to the solenoid value that fixes in the tank containing base through the relay circuit. Suddenly the solenoid value becomes open and basic reagent mixed with the solution. If the PH values of the solution becomes equal to the set point value, then the solenoid valves becomes

* **Corresponding author S. Kannadhasan:** Department of Electronics and Communication Engineering, Study World College of Engineering, Coimbatore, Tamil Nadu-641105, India; E-mail: kannadhasan.ece@gmail.com

S. Kannadhasan, R. Nagarajan, Alagar Karthick, K.K. Saravanan & Kaushik Pal (Eds.)

closed. A slight increase in the PH value of the solution and then the set point value causes the opening of the solenoid valve fixed in the acid reagent tank [1 - 5].

THEORY AND PRINCIPLE

Water is an afunny substance. It makes possible much of the chemistry that goes on in our bodies and all around us. But most people take for granted the chemical properties of water. We've already learned that water molecules are constantly in motion. And keep in mind that each water molecule carries a dipole or net charge across the molecule. As we saw in the atomic bonding lesson, this dipole causes each molecule to behave like a little magnet with a positive and negative end. This dipole causes water molecules to be attracted to each other; the positive hydrogen is attracted to the negative oxygen of a nearby molecule. By PH control, we mean to maintain the PH value during continuous operation at a specific desired value through manipulating the alkaline flow rate. Usually in most industrial applications, the desired value is chosen to be around 7. This is the safest value for portable water, utility water used in industry, or waste disposing water [6-10].

In essence, a PH control system measures the PH of the solution, and controls the addition of a neutralizing agent (on demand) to maintain the solution at the PH of neutrality, or within certain acceptable limits. It is, in effect, a continuous titration. These PH control systems are highly varied, and the design depends on such factors as flow, acid or base strength or variability of strength, the method of adding a neutralizing agent, the accuracy of control (*i.e.*, limits to which pH must be held 0 and physical and other requirements [11-15].

INTRODUCTION TO EMBEDDED SYSTEMS

The embedded system is any electronic equipment but with intelligence and dedicated software. All embedded systems use either a microprocessor or a microcontroller. The application of these microcontrollers makes user- friendly cheaper solutions and enables to add features otherwise impossible to provide by other means. Embedded devices can be defined as any device with an embedded microprocessor or a microcontroller .The software for the Embedded System is called firmware. The firmware is written in Assembly language for time or resource-critical operations or in high-level languages like C or Embedded C. The software will be simultaneously do microcode simulation for the largest processor. Since they are supported to perform only specific tasks, these programs are stored in Read Memory (ROM). Moreover, they may need to have minimal input from the user, hence the user interface like monitor, mouse and large keyboard, *etc.* may be absent [16-18].

Embedded Systems are also known as Real Time systems since they respond to an input or event and produce the result within the guaranteed period. This time period can be a few microseconds to days or months.

Embedded System development: In the development of an Embedded System application, the hardware and software must go hand in hand. The software created by software engineers must be fed into or micro-coded into the hardware or microcontroller produced by the VLSI engineers. The microcontroller and the software micro-coded unit together form the system for a particular application. The Software program for the real-time system is written either in Assembly or a high-level language such as C. The assembly language is used in the case of some critical applications. Nowadays high-level languages replace most of the assembly language constructs.

Embedded technology is present in almost every electronic device we use today (Fig. **1**). There is embedded software inside the cellular phone, automobiles and thermostats in air conditioners, industrial control equipment and scientific and medical equipment, defence use, communication satellites, *etc*. An embedded technology thus covers a broad range of products the generalization of which is difficult.

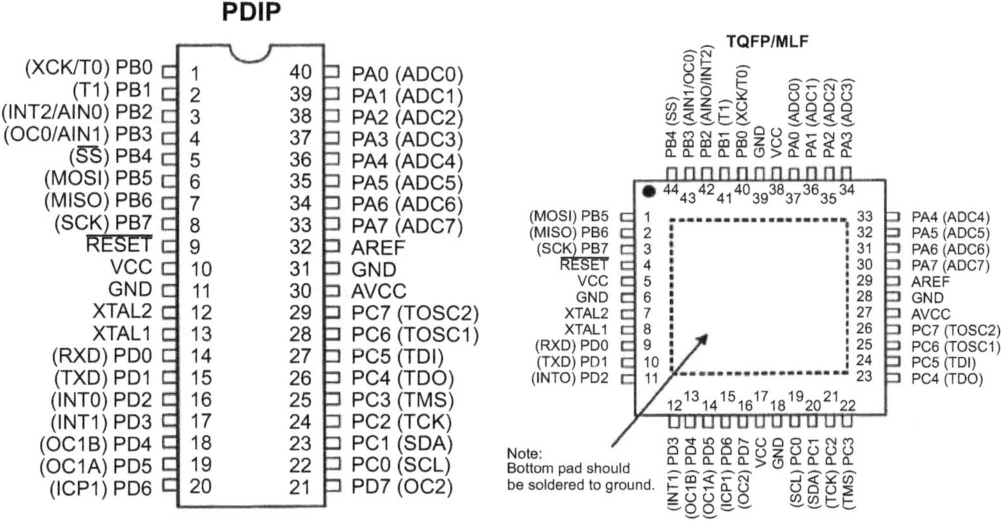

Fig. (1). PIN diagram of ATMEGA 32.

The AVR is a modified hardware architecture 8-bit RISC single-chip microcontroller that was developed by Atmel. The AVR was one of the first microcontroller families to use on-chip flash memory for program storage as opposed to one-time programmable ROM, EPROM, or EEPROM used by other microcontrollers at the time. The Atmel AVR Core combines a rich instruction set with 32 general-purpose working registers. All 32 registers are directly connected to the Arithmetic Logic Unit (ALU), allowing two independent registers to be accessed in one single instruction executed in one clock cycle. The resulting architecture is more code efficient while achieving throughputs up to ten times faster than conventional CISC microcontrollers.

The ATmega32 provides the following features: 32Kbytes of In-system Programmable Flash Program memory with Read-While-Write capabilities, 1024bytes EEPROM, 2Kbyte SRAM, 32 general purpose 1/O lines, 32 general purpose working registers, a JTAG interface for boundary scan, on-chip debugging support and programming, three flexible timer/counters with comparable modes, internal and external interrupts, a serial programmable USART, a byte-oriented Two-wire Serial Interface, an 8- channel, 10-bit ADC with optional differential input stage with programmable gain (TQFP package only) a programmable watchdog timer with internal oscillator, an SPI serial port, and six software selectable power saving modes. The Idle mode stops the CPU while allowing the USART, Two-wire interface, A/D Converter, SRAM; Timer/Counters, SPI port, and interrupt system to continue functioning. The Power-down mode saves the register contents but freezes the oscillator, disabling all other chip functions until the next external interrupt or hardware reset. In the power saving mode, the Asynchronous Timer continues to run, allowing the user to maintain a timer base when the rest of the device is sleeping. The ADC Noise Reduction mode stops the CPU and all I/O modules except Asynchronous Timer and ADC, to minimize the switching noise during ADC conversion. In the standby mode, the crystal/resonator oscillator is running while the rest of the device is sleeping (Fig. **2**). This allows a very fast start-up combined with low-power consumption. In the extended standby mode, both the main oscillator and the asynchronous timer continue to run.

The device is manufactured using Atmel's high-density nonvolatile memory technology. The on-chip ISP flash allows the program memory to be reprogrammed in the system through an SPI serial interface, by a conventional nonvolatile memory programmer, or by an on-chip boot program running on the AVR core. The boot program can use any interface to download the application program from the flash memory. The software in the Boot Flash section will continue to run while the application of flash section is updated, providing a true Read Write operation. By combining an 8-bit RISC CPU with the In System Self Programmable Flash on a monolithic chip, the Atmel Atmega32 is a powerful microcontroller that provides a highly flexible and cost-effective solution to many embedded control applications.

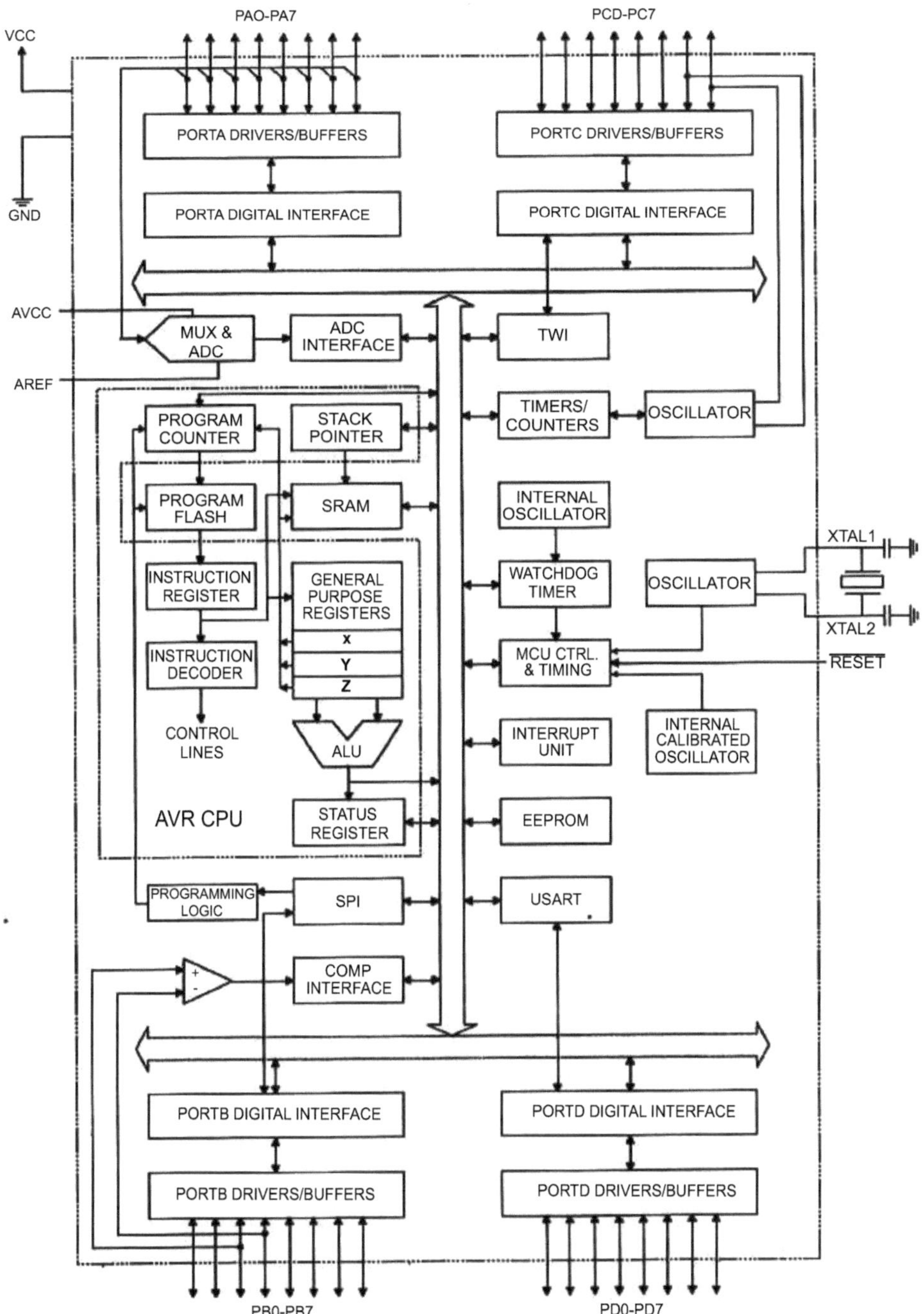

Fig. (2). Block diagram of ATMEGA AVR 32.

The main function of the CPU core is to ensure correct program execution .The CPU must therefore be able to access memory, perform calculations, control peripherals, and handle interrupts. In order to maximize performance and parallelism, the AVR uses a Harvard architecture- with separate memories and buses for programs and data. Instructions in the program memory are executed with a single level pipeline. While one instruction is being executed, the next instruction is pre-fetched from the program memory. This concept enables instructions to be executed in every clock cycle. The program memory is In-System Reprogrammable Flash memory.

PH CONTROL SYSTEMS WORKING

In essence, a PH control system measures the PH of the solution and controls the addition of a neutralizing agent (on demand) to maintain the solution at the PH of neutrality, or within certain acceptable limits. It is, in effect, a continuous titration. These PH control systems are highly varied, and the design depends on factors such as flow, acid or base strength or variability of strength, method of adding a neutralizing agent, the accuracy of control (*i.e.,* limits to which PH must be held), and physical and other requirements. Fig. (**3**) shows the block diagram of the PH control system. In this system, we need 3 main tanks for operation. The first tank is the main tank containing water having an unknown PH value. This PH value is to be controlled to 7. The second and third tanks contain the reagents to neutralize the pH value of water in the first tank. The reagents are acid and base. A PH electrode is dipped in the waste water tank to sense the pH of the water. The output of the electrode is connected to a high gain amplifier circuit containing an op-amp and a I mega ohm resistor. This op-amp circuit amplifies the output of the pH electrode to a desired value. Then the amplified output is transferred to the AVR ATmega32 microcontroller. The program is already written in the microcontroller, the output of AVR is connected to the relay logic circuit.

The working of this PH controlling system is as follows: Firstly the pH electrode senses the pH value of the solution that transfers it to the AVR. In the AVR microcontroller, we can set a desired set point value. If the output of the pH electrode is less than the set point value, and the behaviour of the solution is acidic, then the AVR can compare the pH value of the solution with the set point value. Then the AVR transmits an electrical signal to the solenoid valve fixed in the tank containing the base through the relay circuit. Suddenly, the solenoid valves get opened and the basic reagent is mixed with the solution. If the pH value of the solution becomes equal to the set point value, then the solenoid valve becomes closed. A slight increase in the pH value of the solution than the set point value causes the opening of the solenoid valve fixed in the acid reagent tank.

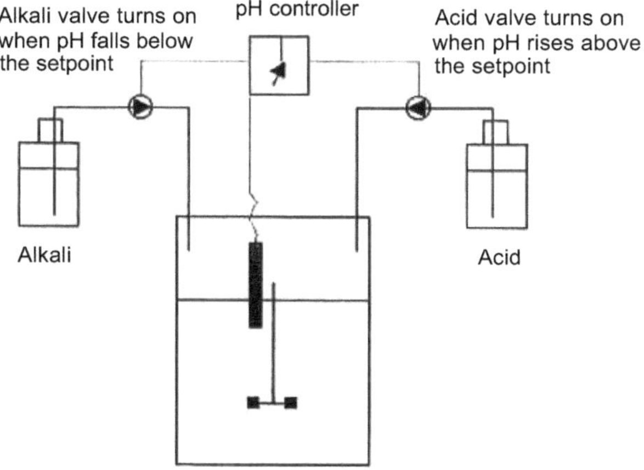

Fig. (3). Block diagram of PH controller.

ZIGBEE 3.0 PROTOCOL

Zigbee is a wireless technology developed as an open global standard to address the unique needs of low-cost, low-power wireless IoT networks. The Zigbee standard operates with the IEEE 802.15.4 physical radio specification and operates in unlicensed bands including 2.4 GHz, 900 MHz, and 868 MHz. The 802.15.4 specification, upon which the Zigbee stack operates, gained ratification by the Institute of Electrical and Electronics Engineers (IEEE) in 2003. The specification is a packet-based radio protocol intended for low-cost, battery-operated devices (Fig. **4**). The protocol allows devices to communicate in a variety of network topologies and can have a battery life lasting several years. The Zigbee 3.0 protocol was created and ratified by member companies of the Zigbee Alliance. Over 300 leading semiconductor manufacturers, technology firms, OEMs, and service companies comprise the Zigbee Alliance membership.

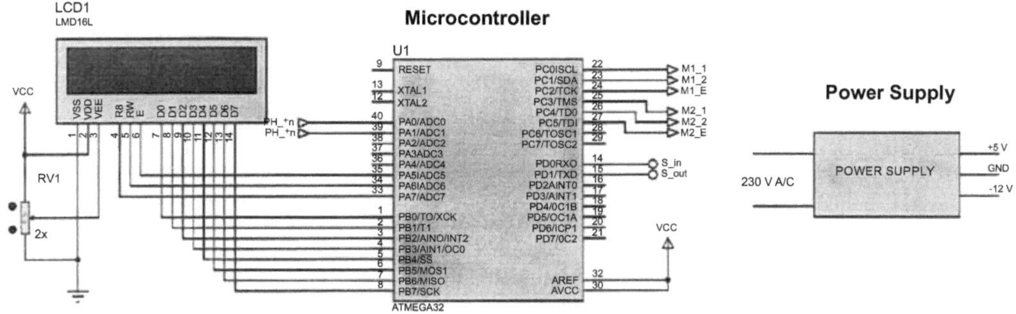

Fig. (4). Circuit arrangement of PH monitor.

The Zigbee protocol was designed to provide an easy-to-use wireless data solution characterized by secure, reliable wireless network architectures. The Zigbee 3.0 protocol is designed to communicate data through noisy RF environments that are common in commercial and industrial applications. Version 3.0 builds on the existing Zigbee standard but unifies the market-specific application profiles to allow all devices to be wirelessly connected in the same network, irrespective of their market designation and function. Additionally, a Zigbee 3.0 certification scheme ensures the interoperability of products from different manufacturers. Connecting Zigbee 3.0 networks to the IP domain opens up monitoring and control from devices such as smartphones and tablets on a LAN or WAN, including the Internet, and brings the true Internet of Things to fruition.

Fig. (**5**) depicts a Zigbee 3.0 protocol stack. As shown, it consists of PHY, MAC, network, and application layers. The changes have been incorporated into the application layer in Zigbee 3.0 compared to previous Zigbee versions. The network layer is sand-witched between the upper layer, *i.e.,* the application layer, and PHY/MAC as defined in IEEE 802.15.4 standard. The Zigbee 3.0 protocol stack incorporates Zigbee-based device layer which provides consistent behaviour for commissioning new nodes in the network. The Zigbee 3.0 protocol is designed to communicate data through noisy RF environments that are common in commercial and industrial applications. Version 3.0 builds on the existing Zigbee standard but unifies the market-specific application profiles to allow all devices to be wirelessly connected in the same network, irrespective of their market designation and function. Additionally, a Zigbee 3.0 certification scheme ensures the interoperability of products from different manufacturers. Connecting Zigbee 3.0 networks to the IP domain opens up monitoring and control from devices such as smartphones and tablets on a LAN or WAN, including the Internet, and brings the true Internet of Things to fruition.

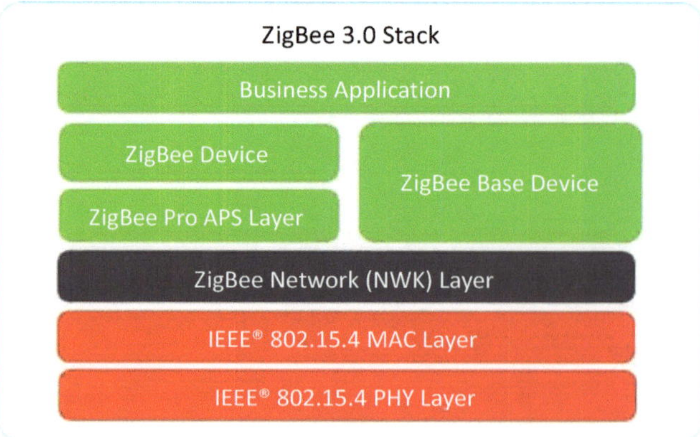

Fig. (5). ZigBee 3.0 Protocol stack.

WIRELESS PH MONITORING USING ZIGBEE 3.0 PROTOCOL

A sensor unit basically consists of several sensors used to detect the predetermined parameters that indicate the quality of water. In this work, three types of sensors are mentioned; a PH sensor that senses the acidity or basicity of the water, a temperature sensor, and a turbidity sensor based on a phototransistor are used. All sensors use a battery for their operation. The information being sensed by the sensors is then converted into an electrical signal and sent through the signal conditioning circuit that functions to make sure the voltage or current produced by the sensors is proportional to the actual values of the parameters being sensed. Then it is passed to a microcontroller or a microprocessor that processes it to the value understandable by humans.

The main microprocessor of the sensor node is based on the Zigbee compliance product from Cirronet. The high power transmission type ZMN2405HP Zigbee module uses the CC2430 transceiver IC from Texas Instrument complying with the IEEE 802.15.4 standard with a maximum transmission power of 100 mW using the dipole antenna, and 250 mW using the directional patch antenna. The transceiver IC is integrated with the 8051 microcontroller with a low power but high performance of 64 kByte programmable flash features. The module alone requires a 5VDC power supply, multiple sensor inputs/output with ADC, operating at a frequency of 2.4 GHz with a configurable sleep mode to get the best of power consumption as low as 3uA. The main microcontroller in the module is reprogrammable whether to function as an end device, router, or coordinator node. As an end device sensor node, it can only communicate with the router or coordinator to pass the data to the sensor. An end device can only communicate indirectly with the other end device through the router or coordinator. The sensor node, defined as a router, is responsible for routing data from other routers or the end devices to the coordinator or to other routers closer to the coordinator. The router can also be a data input device like the end device, but in actual application, it is generally used to extend the coverage distance of the monitoring system. There can be only one coordinator for the monitoring system. The coordinator is responsible for setting the channel for the network to use, assigning the network address to routers and end devices, and keeping the routing tables for the network that are necessary to route data from one end device to another in the same Zigbee network (Fig. **6**).

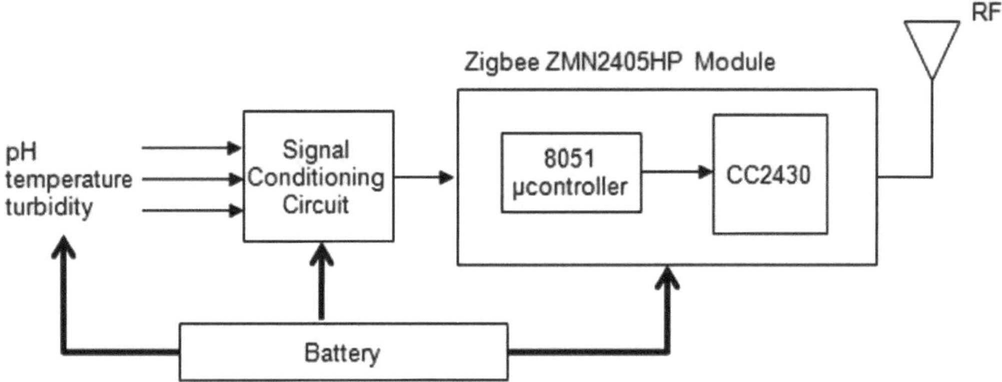

Fig. (6). Block diagram of Zigbee based wireless PH monitoring system.

CONCLUSION

The pH control is quite difficult due to the nonlinearity of the neutralization process. This process requires good control to overcome load disturbances. This paper has demonstrated how a Genetic Algorithm based pH process can be used for optimum control. The complete methodology is presented, focusing on each part of the adaptive system. Experiments show that it is possible to control the pH value to 7 using atmega32. A pH controller can be made at low cost with AVR. Because of the use of the microcontroller, it is possible to maintain the system by changing the program and we can also upgrade it. This system gives the solution to the industrial wastewater problem.

REFERENCES

[1] M.J. Chae, H.S. Yoo, J.R. Kim, and M.Y. Cho, "Bridge Condition Monitoring System Using Wireless Network (CDMA and Zigbee)", *23rd International Symposium on Automation and Robotics in Construction ISARC,* Tokyo, Japan, 3 – 5 Oct, 2006.
[http://dx.doi.org/10.22260/ISARC2006/0064]

[2] Sangmi Shim, Seungwoo Park, and Seunghong Hong, "Parking Management System Using Zigbee", *International Journal of Computer Science and Network Security,* vol. 6, no. 9B, 2006.

[3] F. Vergari, V. Auteri, C. Corsi, and C. Lamberti, "A Zigbee-based ECG Transmission For Low Cost Solution In Home Care Services Delivery", *Mediterranean Journal of Pacing and Electrophysiology,* 2009.http:// www.mespe.net/en/newselem

[4] "Cirronet, ZMN2405/HP ZigbeeTM Module Developer's Kit User Manual", *Rev A,* 2007.

[5] "Wireless Medium Access Control (MAC) and Physical Layer (PHY) Specifications for Low Rate Wireless Personal Area Network (LR-WPANs)", *IEEE Standard 802.15.4 TM,* 2003

[6] J-S. Lee, Y-W. Su, and C-C. Shen, "A comparative study of wireless protocols: Bluetooth, UWB, ZigBee, and Wi-Fi", *IECON 2007-33rd Annual Conference of the IEEE Industrial Electronics Society,* pp.46-51, 2007.
[http://dx.doi.org/10.1109/IECON.2007.4460126]

[7] B.E. Horton, S. Schweitzer, A.J. DeRouin, and K.G. Ong, "A varactor-based inductively coupled wireless pH sensor", *IEEE Sens. J.,* vol. 11, no. 4, pp. 1061-1066, 2011.

[http://dx.doi.org/10.1109/JSEN.2010.2062503]

[8]   S. Bhadra, G. E. Bridges, D. J. Thomson, and M. S. Freund, "A wireless passive pH sensor based on pH electrode potential measurement", *Sensors,* pp. 927-930, 2010.
[http://dx.doi.org/10.1109/ICSENS.2010.5690888]

[9]   https://www.digi.com/blog/post/understanding-the-zigbee-3-0-protocol

[10]   J.B. Ong, Z. You, J. Mills-Beale, E.L. Tan, B.D. Pereles, and K.G. Ong, "A wireless passive embedded sensor for real time monitoring of water content in civil engineering materials", *IEEE Sens. J.,* vol. 8, no. 12, pp. 2053-2058, 2008.
[http://dx.doi.org/10.1109/JSEN.2008.2007681]

[11]   J.J. Barron, C. Ashton, and L. Geary, "The effects of temperature on pH measurement", *A Reagecon Technical Paper,* 2005.

[12]   S. Bhadra, G.E. Bridges, D.J. Thomson, and M.S. Freund, "A wireless passive sensor for pH monitoring employing temperature compensation", *SENSORS,* pp. 1522-1525, 2011.
[http://dx.doi.org/10.1109/ICSENS.2011.6127181]

[13]   A.N. Das, F.L. Lewis, and D.O. Popa, "Data-Logging and Supervisory Control in Wireless Sensor Networks", *Proceedings of the Seventh ACIS International Conference on Software Engineering, Artificial Intelligence, Networking, and Parallel/Distributed Computing (SNPD'06),.* June 19- 20, pp. 330-380, 2006,
[http://dx.doi.org/10.1109/SNPD-SAWN.2006.28]

[14]   Q. Wang, and Y. Wang, "Design and implementation of smart home remote monitoring system based on ARM11", *2018 Chinese Control And Decision Conference (CCDC),* pp. 209-213, 2018.
[http://dx.doi.org/10.1109/CCDC.2018.8407132]

[15]   R.L. Levine, R. Fromm Jr, M. Mojtahedzadeh, A.A. Baghaje, and A.R. Opekun Jr, "Equivalence of litmus paper and intragastric pH probes for intragastric pH monitoring in the intensive care unit", *Crit. Care Med.,* vol. 22, no. 6, pp. 945-948, 1994.
[http://dx.doi.org/10.1097/00003246-199406000-00011] [PMID: 7911416]

[16]   D. Meiners, S. Clift, and D. Kaminski, "Evaluation of various techniques to Monitor Intragastric pH", *Arch Surg,* vol. 117, no. 3, pp. 288-291, 1982.
[http://dx.doi.org/10.1001/archsurg.1982.01380270016004]

[17]   A.A. Baghaie, M. Mojtahedzadeh, R.L. Levine, N. Fromm Robert E Jr, K.K. Guntupalli, and A.J. Opekun Jr, "Comparison of the effect of intermittent administration and continuous infusion of famotidine on gastric pH in critically ill patients", *Crit. Care Med.,* vol. 23, no. 4, pp. 687-691, 1995.
[http://dx.doi.org/10.1097/00003246-199504000-00017] [PMID: 7712759]

[18]   G. Mois, S. Folea, and T. Sanislav, "Analysis of three IoT-based wireless sensors for environmental monitoring", *IEEE Trans. Instrum. Meas.,* vol. 66, no. 8, pp. 2056-2064, 2017.
[http://dx.doi.org/10.1109/TIM.2017.2677619]

CHAPTER 6

Identification OF Differential Pattern in ADES AL: Initiation

Harishchander Anandaram[1,*]

[1] *Centre for Excellence in Computational Engineering and Networking, Amrita Vishwa Vidyapeetham, Coimbatore, Tamil Nadu, India*

Abstract: *Aedes albopictus* is considered the primary threatening vector affecting public health. The process of identifying the specific transcripts for enhancing the growth factor in *Aedes albopictus* is the initiation towards the development of a therapeutic marker. It implicates the identification of a particular antagonist. The approach was a reference-based analysis of the whole transcriptome to reveal the differentially expressed pattern of transcripts. Further research requires the mathematical modeling of gene regulation and differential expression.

Keywords: ADES, AL, Dengue, Mapping.

INTRODUCTION

The *albopictus* species of *Aedes* is an essential vector for various human pathogens, including Zika, chikungunya, and dengue viruses. *Aedes albopictus a* is a competent vector of medical importance with a lack of reliable information on essential traits. Strategies for controlling the *albopictus* species of *Aedes* mosquitoes are limited [1-3]. This species of Aedes has a large genome with significant population-based size variation. Aedes mosquitoes transmit chikungunya. In recent years, various studies have determined the competence of *Aedes albopictus* for Zika. Understanding the mechanism of gene regulation associated with insecticide resistance is vital to developing a novel diagnostic biomarker and the differential gene expression of deltamethrine in *ae. Albopictus* from various studies incorporating the RNAseq technology [4-6].

---

* **Corresponding author Harishchander Anandaram:** Centre for Excellence in Computational Engineering and Networking, Amrita Vishwa Vidyapeetham, Coimbatore, Tamil Nadu, India; E-mail: a_harishchander@cb.amrita.edu

MATERIALS AND METHODS

Analytics

The data analysis of the transcript in the platform for sequencing RNA detects multiple patterns of noncoding RNA.

Experiment

Transcriptome sequencing experiments include RNA extraction and Quality Control (QC), and each step is essential to ensure data quality.

Mapping

It is necessary to choose the reference ordination for optimum analysis. Sequences at intervals in the deoxyribonucleic acid region are additionally attributable to immature RNA contamination or incomplete genomic annotation. In contrast, sequences mapped to the intergenic region are additionally attributable to incomplete genomic annotations and background.

Quantification

The approach involves the quantification of the number of reads that match the reference genome (read count). This quantification in the sequence level uses a standard sequence format. Each model represents the structure of transcripts created by a given sequence, and the live TPM (Transcript per million reads) accounts for every sequence length and library size.

Expression

The level of differential gene expression is studied based on "read density"; *i.e.,* read density is directly proportional to genes' expression level—application of Salmon TPM (Transcripts Per Million reads). The read count mapped to the gene normalized to total read counts.

Annotation

Gene Set Enrichment Analysis (GSEA) involves the Functional Annotation (FA) of the Differentially Expressed Genes (DEG).

RESULT

After Executing (a) quality control, (b) mapping, (c) quantification, and (d) differential gene expression analysis, we get the results containing differentially.

expressed genes, *i.e.,* co-expressed, up-regulated, and down-regulated. The red color region represents a high expression rate in the heat map, and the blue area represents a low expression rate Fig. (**1**).

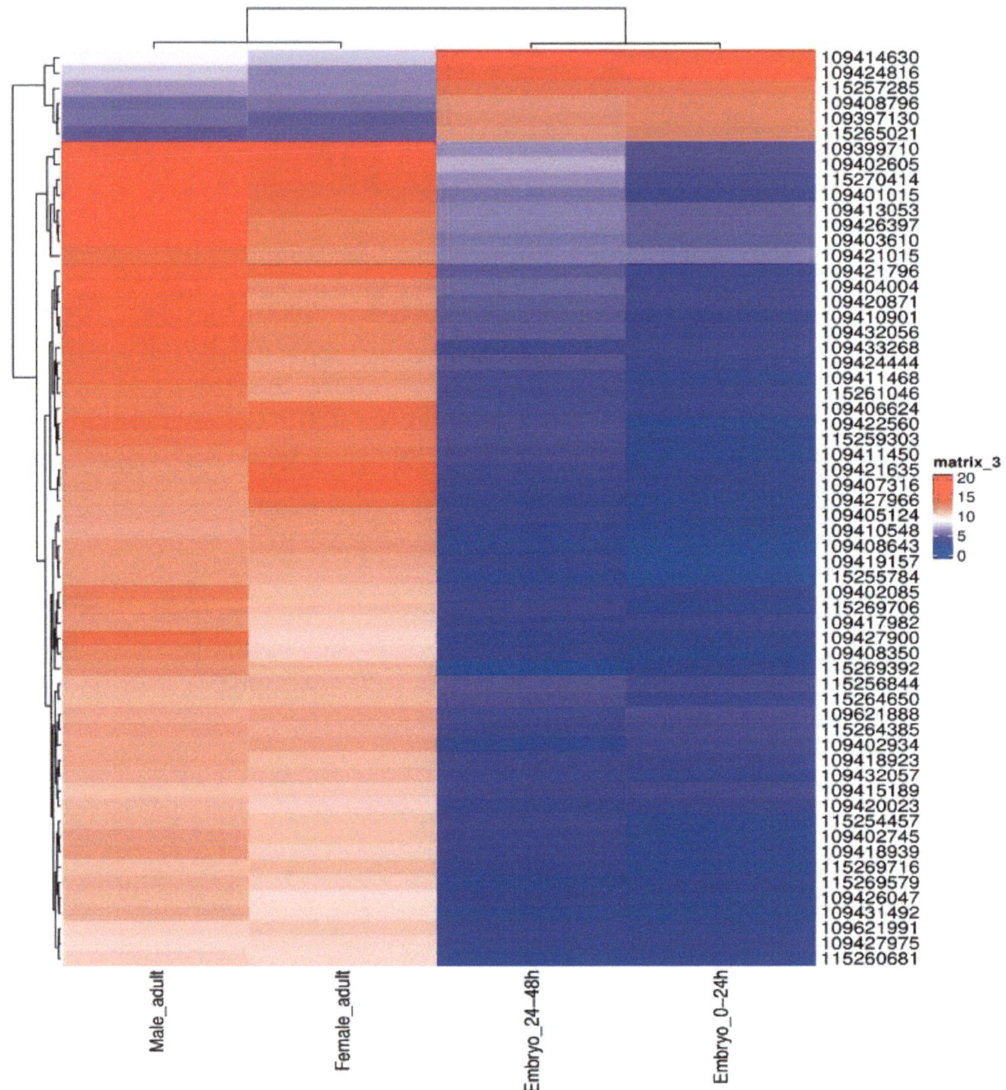

Fig. (1). Heat map (Embryo_*vs*_Adult).

The volcano plot shows differentially expressed regions in red color and baseline regions in green color. Differentially expressed genes are segregated based on the log 2 Fold Change (FC) following the p-value. If the log2FC is >= 1, then those genes are categorized as up-regulated genes, and if the p-value is <= -1, those genes are categorized as down-regulated genes. If genes come in between -1 and 1, those regions as baseline genes Fig. (**2**).

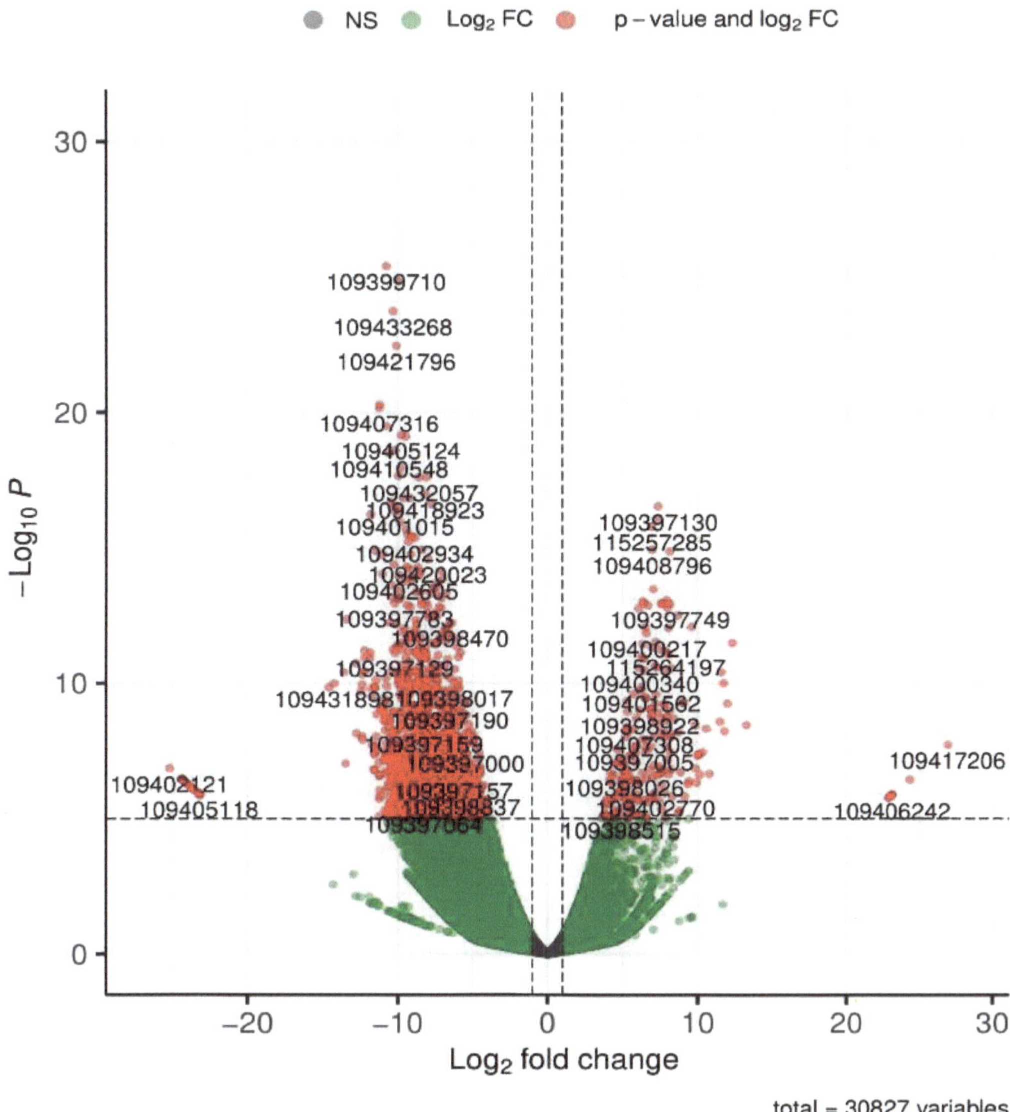

Fig. (2). Volcano plot: (Embryo_*vs*_Adult).

The image below shows that most larvae transcripts are expressed higher than embryo transcripts, except a few at the bottom that follow opposite patterns Fig. (3).

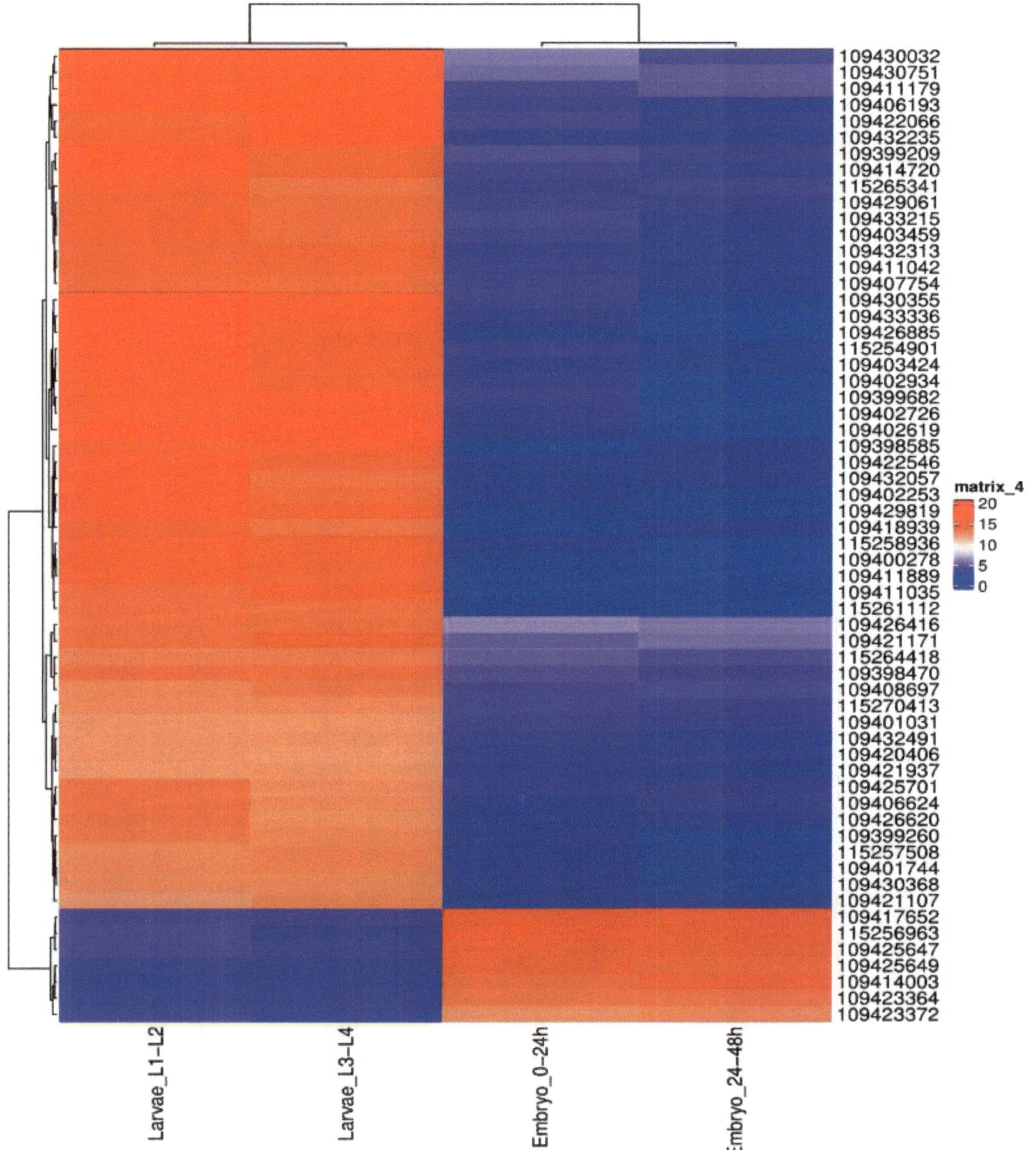

Fig. (3). Heat map: (Larvae_*vs*_Embryo).

The below volcano plot indicates that out of more than 30000 transcripts, there may be a bright chance to determine the growth-enhancing genes in these stages Fig. **(4)**.

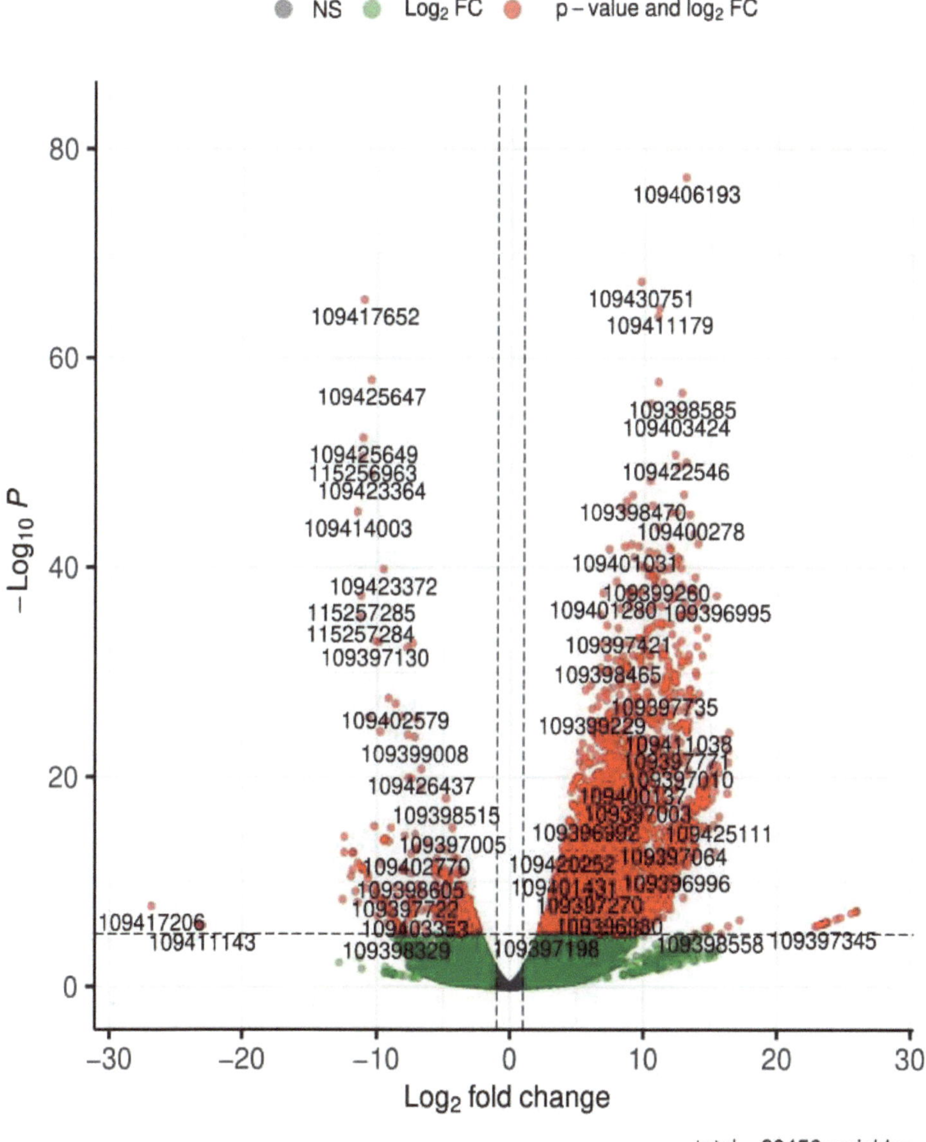

Fig. (4). Volcano Plot: (Larvae_*vs*_Embryo).

The below heat map shows that pupae transcripts have a higher expression rate than an embryo. A small number of transcripts in the source expressed higher than pupae Fig. (**5**).

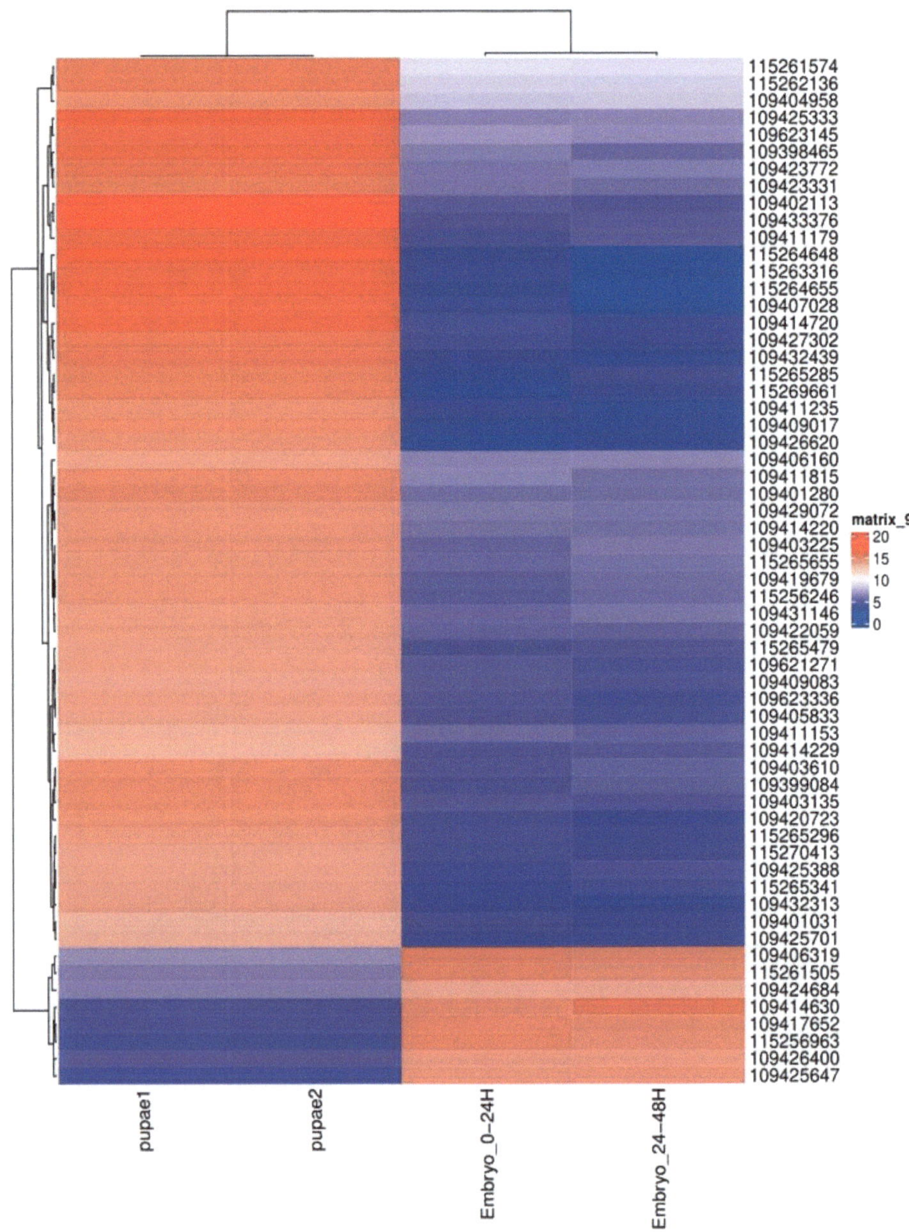

Fig. (5). Heat map: (Pupae_*vs*_Embryo).

The below volcano plots explain that differentially expressed transcripts (both up- and down-regulated) are higher than the baseline. Here also we may expect growth-enhancing genes in these two stages Fig. (**6**).

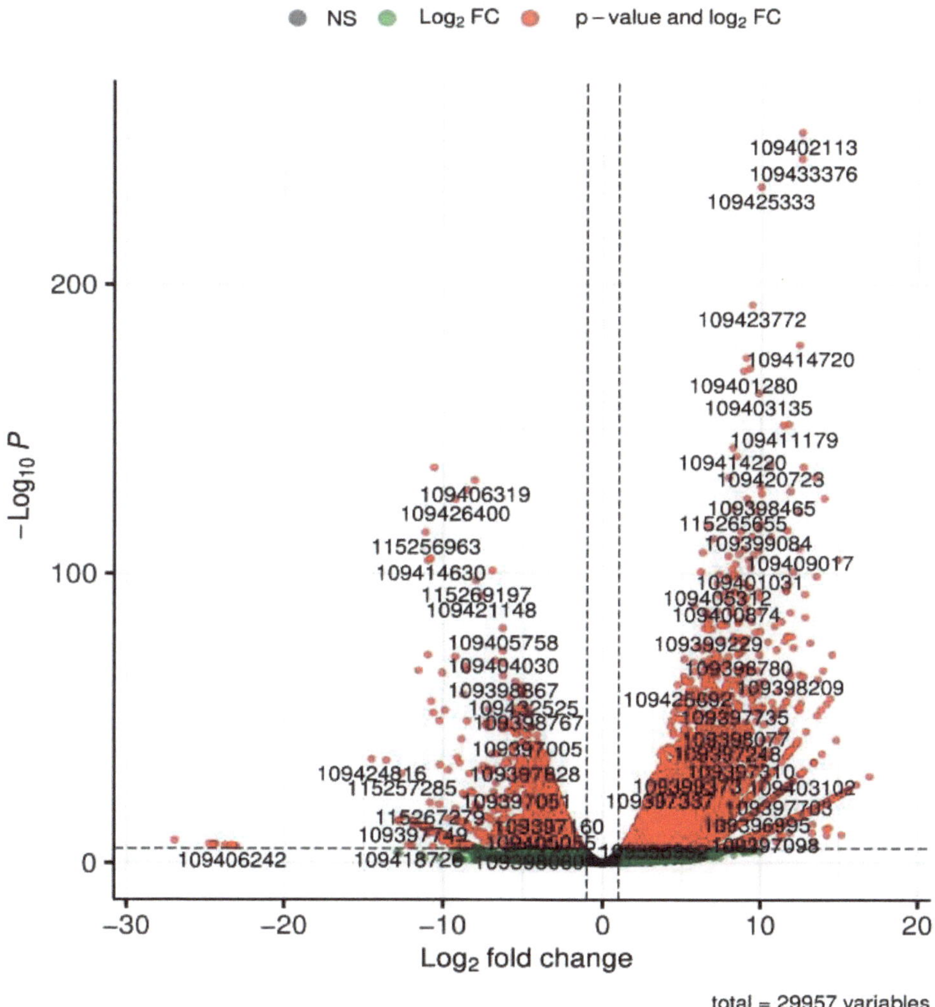

Fig. (6). Volcano Plot: (Pupae_*vs*_Embryo).

The below image shows that female adult transcripts have a higher expression rate than male adult transcripts Fig. (**7**).

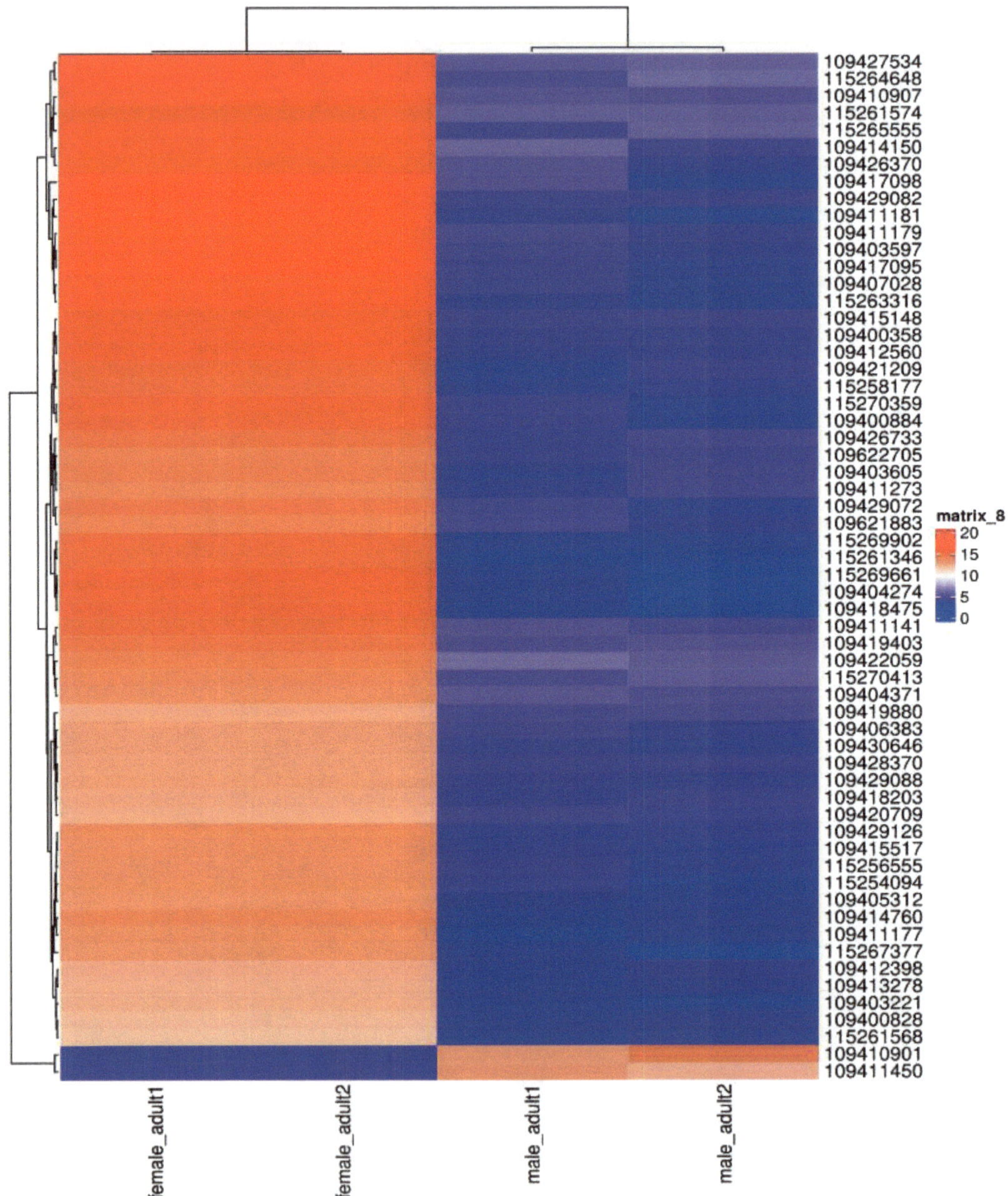

Fig. (7). Heat map: (Female_*vs*_Male).

The below-mentioned plot indicates a differentially expressed pattern for a few transcripts meaning that there will not be many expressions once a mosquito develops into an adult Fig. (**8**).

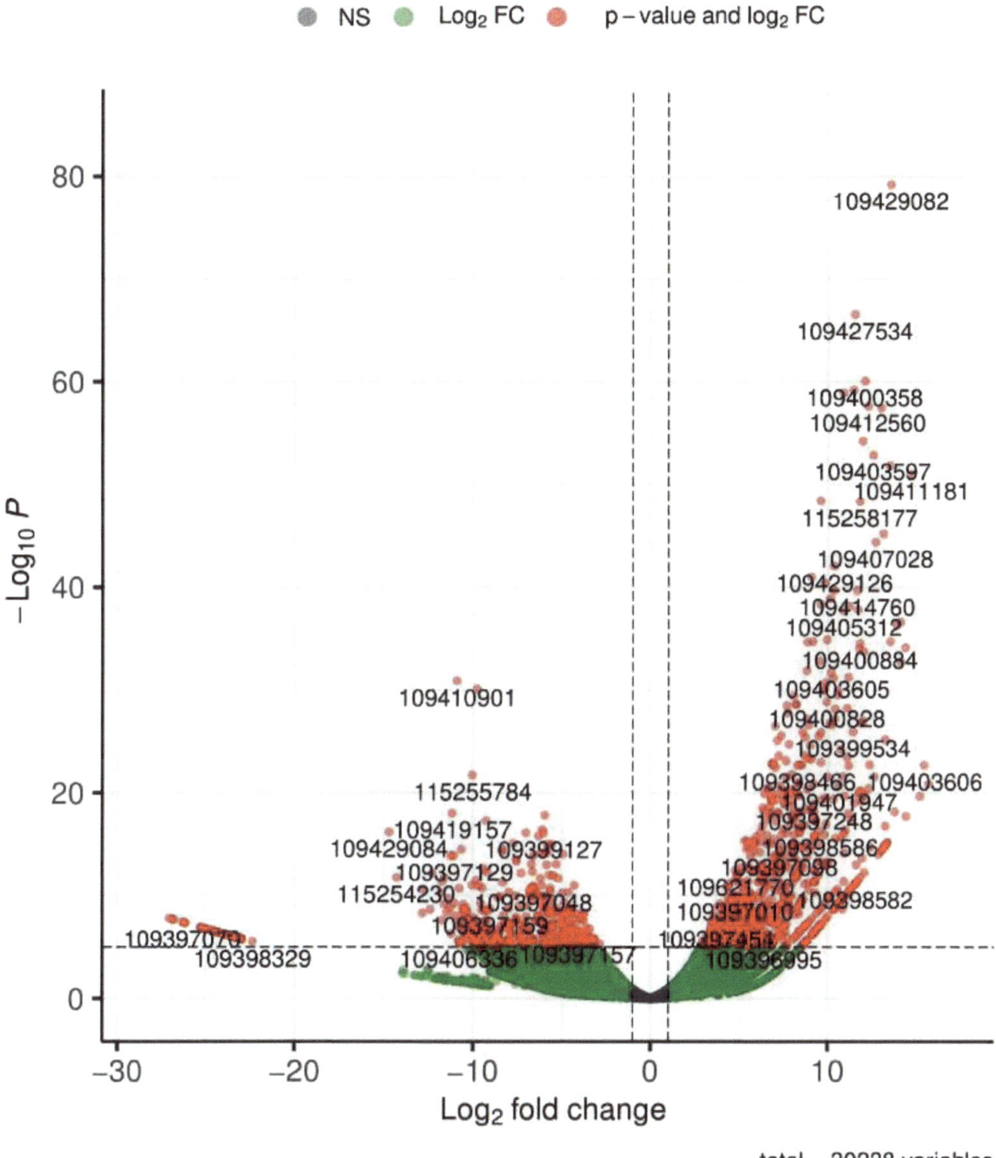

Fig. (8). Volcano Plot: (Female_*vs*_Male).

The below heat map explains that the expression of transcripts in larvae is higher than in adults, and it means there may be a chance that genes express higher during the development stages than in adults Fig. (**9**).

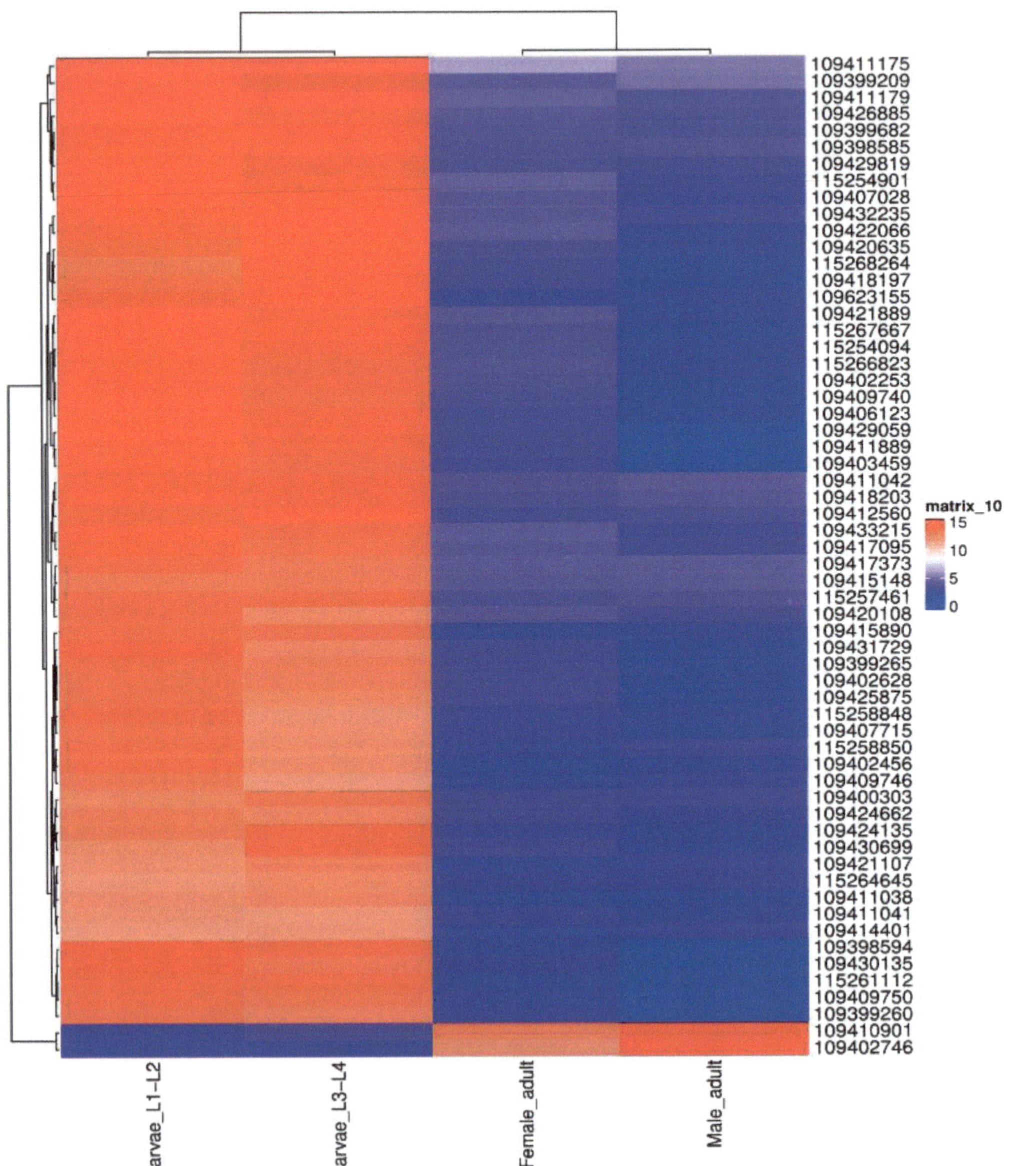

Fig. (9). Heat map: (Larvae_*vs*_Adult).

In the below-mentioned volcano plot, the upregulated transcripts are higher than downregulated out of over 30000 transcripts Fig. (**10**).

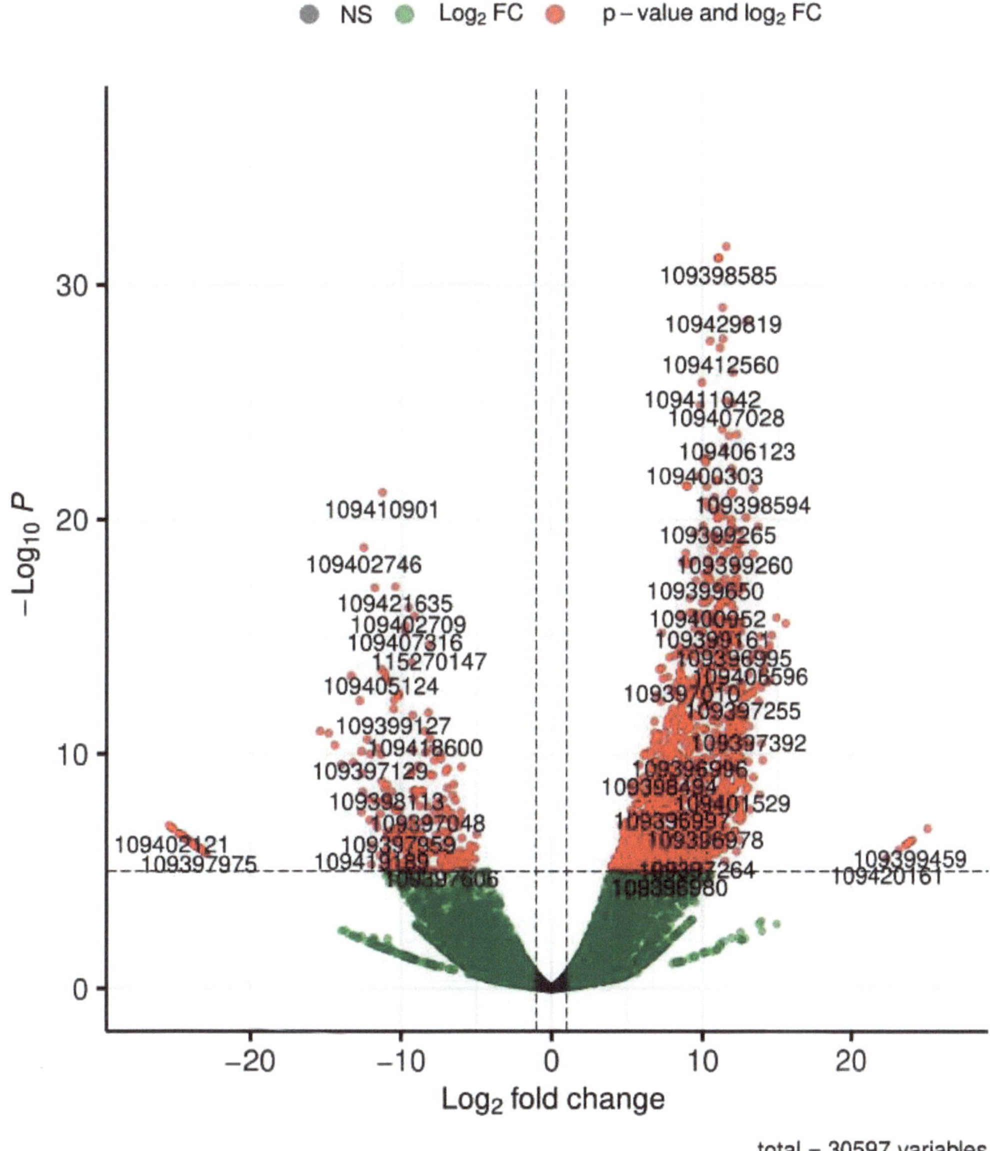

Fig. (10). Volcano Plot: (Larvae_*vs*_Adult).

In general perception, it is noticeable that larvae expressions are much higher than the other stages of the mosquito Fig. (**11**).

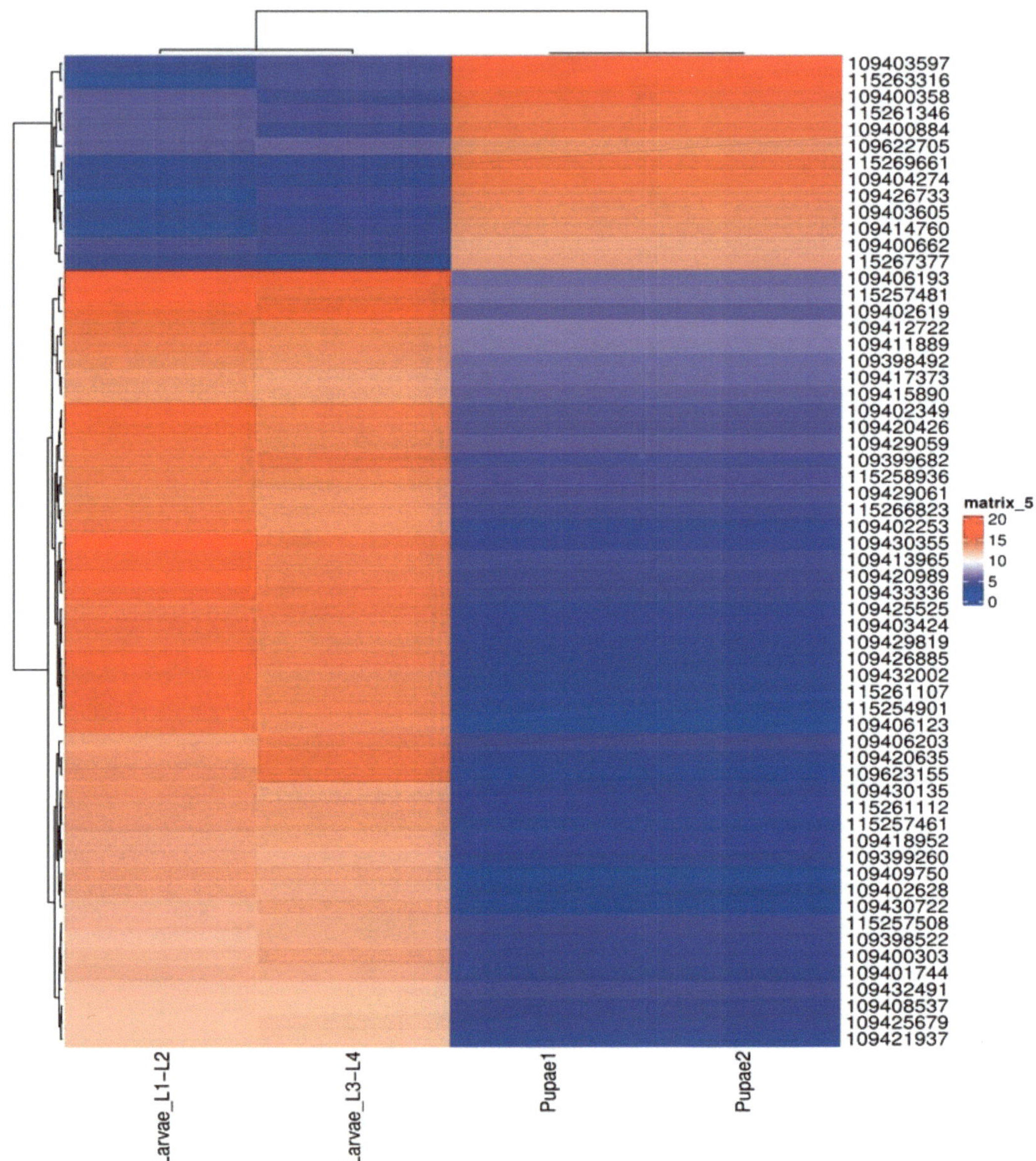

Fig. (11). Heat map: (Larvae_*vs*_Pupae).

In this volcano plot, we can see a significant number of transcripts that are differentially expressed (up-and down-regulated) with a higher number of down-regulated transcripts Fig. (**12**).

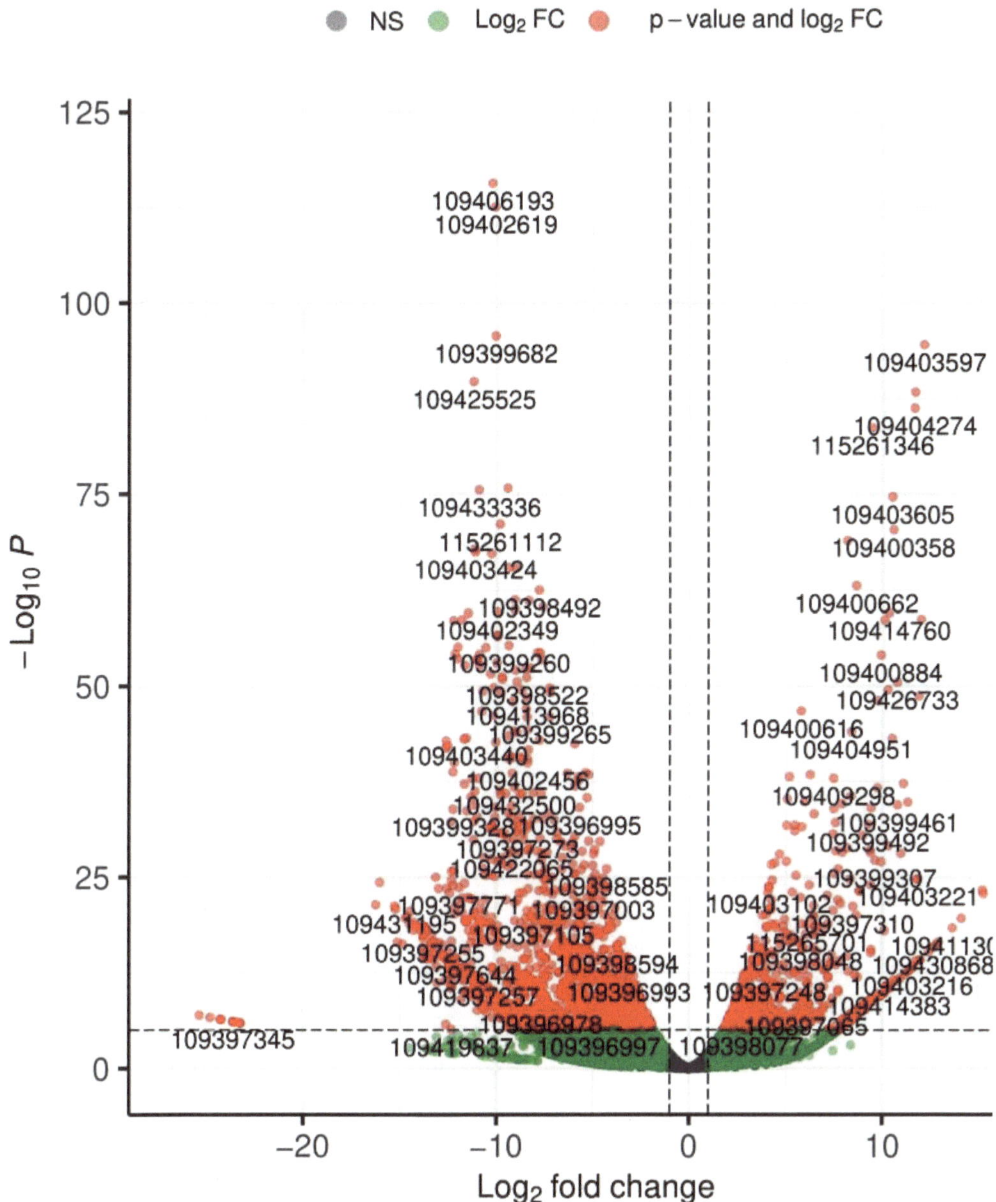

Fig. (12). Volcano Plot: (Larvae_*vs*_Pupae).

CONCLUSION

Mosquitos are the primary vectors carrying infectious pathogens that cause yellow fever, Zika fever, malaria, dengue, and filariasis. The observation of differentially expressed genes in mosquito growth at different developmental stages is the initiative to develop a competent molecule against viruses.

REFERENCES

[1]   V.H. Ferreira-de-Lima, and T.N. Lima-Camara, "Natural vertical transmission of dengue virus in Aedes aegypti and Aedes albopictus: A systematic review", *Parasit. Vectors,* vol. 11, no. 1, p. 77, 2018.
[http://dx.doi.org/10.1186/s13071-018-2643-9] [PMID: 29391071]

[2]   Z. Huang, J. Kim, R.S. Lacruz, P. Bringas Jr, M. Glogauer, T.G. Bromage, V.M. Kaartinen, and M.L. Snead, "Epithelial-specific knockout of the Rac1 gene leads to enamel defects", *Eur. J. Oral Sci.,* vol. 119, no. 0 1, suppl. Suppl. 1, pp. 168-176, 2011.
[http://dx.doi.org/10.1111/j.1600-0722.2011.00904.x] [PMID: 22243243]

[3]   E.B. Kauffman, and L.D. Kramer, "Zika virus mosquito vectors: Competence, biology, and vector control", *J. Infect. Dis.,* vol. 216, suppl. Suppl. 10, pp. S976-S990, 2017.
[http://dx.doi.org/10.1093/infdis/jix405] [PMID: 29267910]

[4]   N. Liu, "Insecticide resistance in mosquitoes: Impact, mechanisms, and research directions", *Annu. Rev. Entomol.,* vol. 60, no. 1, pp. 537-559, 2015.
[http://dx.doi.org/10.1146/annurev-ento-010814-020828] [PMID: 25564745]

[5]   J. Bagi, N. Grisales, R. Corkill, J.C. Morgan, S. N'Falé, W.G. Brogdon, and H. Ranson, "When a discriminating dose assay is not enough: measuring the intensity of insecticide resistance in malaria vectors", *Malar. J.,* vol. 14, no. 1, p. 210, 2015.
[http://dx.doi.org/10.1186/s12936-015-0721-4] [PMID: 25985896]

[6]   G. Benelli, "Research in mosquito control: Current challenges for a brighter future", *Parasitol. Res.,* vol. 114, no. 8, pp. 2801-2805, 2015.
[http://dx.doi.org/10.1007/s00436-015-4586-9] [PMID: 26093499]

CHAPTER 7

Comparative Modelling and Binding Compatibility of Bi-Functional Proteins in Microcystis aeruginosa

Harishchander Anandaram[1,*]

[1] *Centre for Excellence in Computational Engineering and Networking, Amrita Vishwa Vidyapeetham, Coimbatore, Tamil Nadu, India*

Abstract: The objective of the study was to identify a potential inhibitor for Bifunctional Protein in Microcystisaeruginosa. The *in silico* modeling of the protein using the "TBM" module of "Galaxy Seok Lab" extended the execution of virtual screening using MTi open screen. Finally, the protein-ligand interaction was studied using LIGPLOT software for "Bifunctional Protein" in "Microcystis aeruginosa." The virtual screening revealed 7176 compounds from the drug library, and the "best fit" screening resulted in 1500 compounds. Among the 1500 compounds, the molecule MK-3207 showed a better affinity towards the bifunctional Protein with -11.3Kcal/mol binding energy.

Keywords: Aeruginosa, Bi, Blooms, Ligplot, Proteins.

INTRODUCTION

The primary duty of humans in the twenty-first century is to maintain water quality. Mostly, cyanobacteria grow in unsafe water. The species of Phytoplankton species grow in eutrophic water bodies. The most common species of cyanobacteria in freshwater settings that range from tropical to cold subzones is Microcystis aeruginosa. Blooms of *M. aeruginosa* create various environmental issues, including an unpleasant odor, but the most significant problem is developing microcystins.

MATERIALS AND METHODS

Protein Modeling: Template-based modeling (TBM) is a structure prediction approach that uses homologous proteins as templates in this context. Computational Screening of Leads: The web server MTiOpen Screen comprises two services, MTiAutoDock and MTiOpen Screen. MTiAutoDock supports doc-

* **Corresponding author Harishchander Anandaram:** Centre for Excellence in Computational Engineering and Networking, Amrita Vishwa Vidyapeetham, Coimbatore, Tamil Nadu, India; E-mail: a_harishchander@cb.amrita.edu

S. Kannadhasan, R. Nagarajan, Alagar Karthick, K.K. Saravanan & Kaushik Pal (Eds.)

king compounds into a predetermined or user-defined binding site and a rigid docking with Autodock 4.2. MTiOpen Screen automates virtual screening by docking with Autodock Vina.

Analysis of Non Bonding Interactions: Ligplot tool of the European Bioinformatics Institute (EBI) gave the details of H bonding along with Vander Wall and Columb interaction [1-10].

RESULT

The Rampage server's target protein quality provided knowledge of nonbonding interactions between the protein and the ligands (Figs. **1-3**) to determine their binding free energy (Figs. **4-6**). This MTi automated docking research identified 7176 compounds in the drug library and the best fit of 1500 molecules (Figs. **7-9**). The Protein had the most excellent relationship with the Mk3207 chemical, with a -11.3Kcal/mol binding energy. Because of the reduced binding energy, the ligand is more stable than another molecule (Figs. **10-13**).

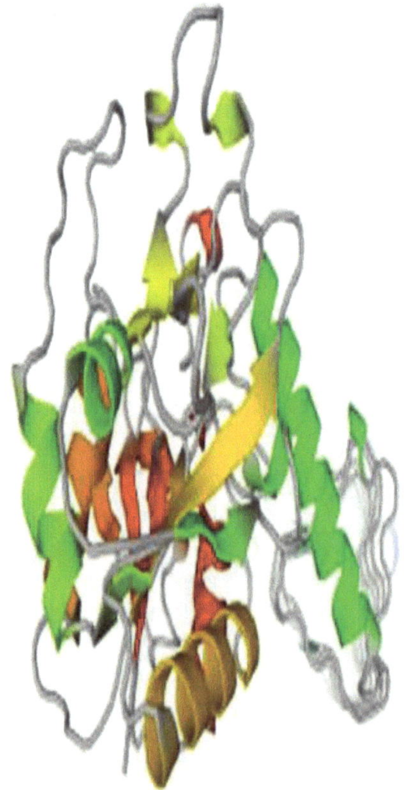

Fig. (1). Predicted Structure of Bifunctional Protein.

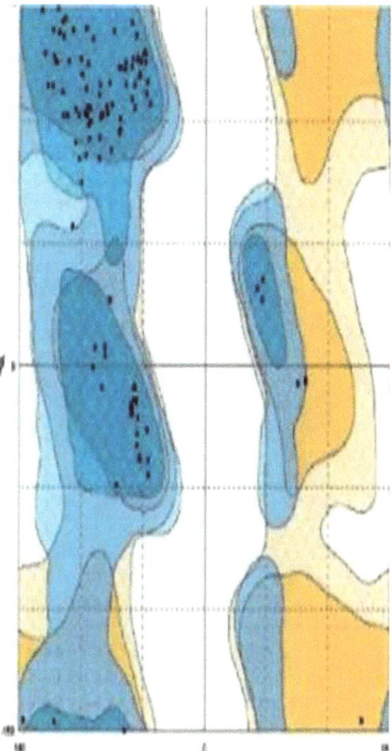

Fig. (2). Quality of Predicted Structure.

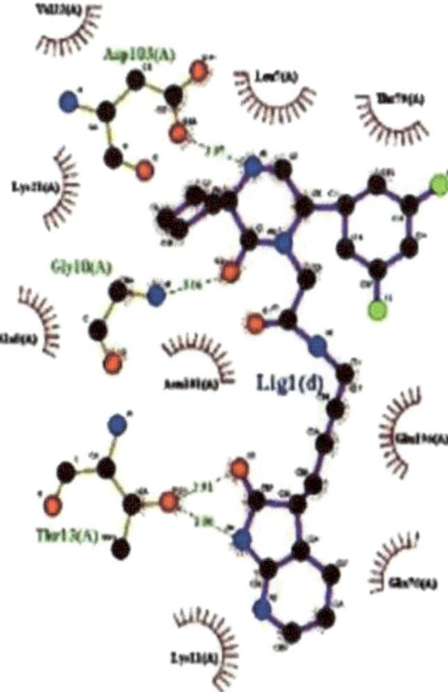

Fig. (3). Protein-Ligand Interaction.

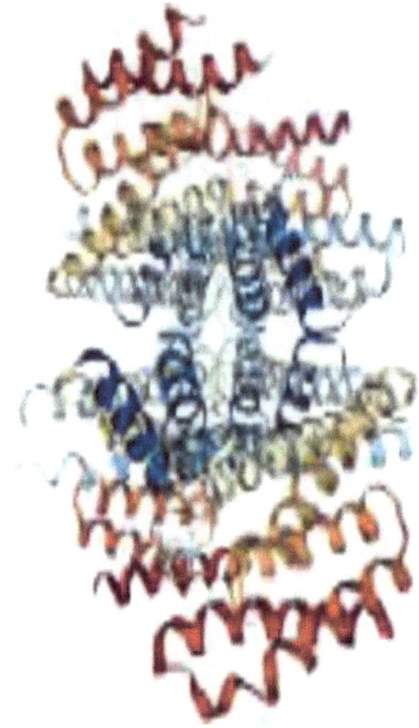

Fig. (4). 14-3-3 Protein.

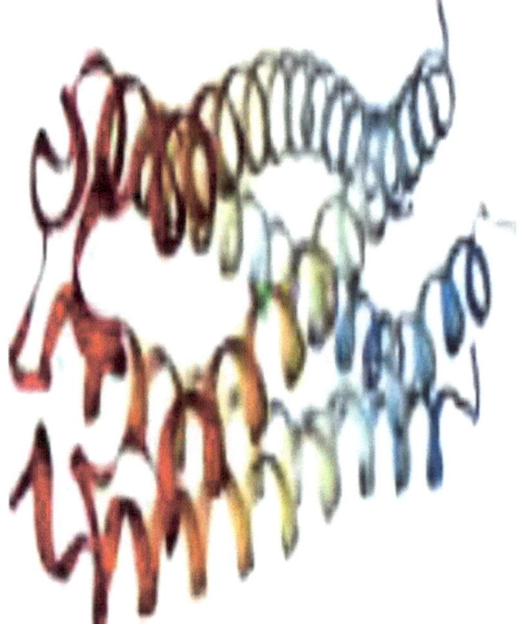

Fig. (5). Cartilage oligomeric Matrix Protein.

Fig. (6). Vcam-1.

Fig. (7). Calreticulin.

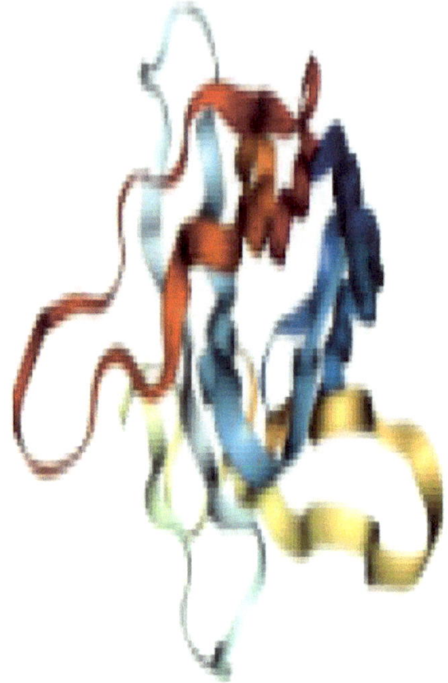

Fig. (8). Gelsolin Nanobody.

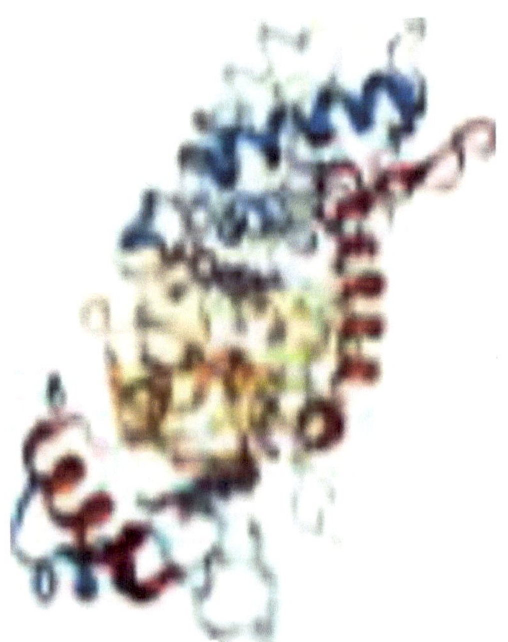

Fig. (9). CCP3.

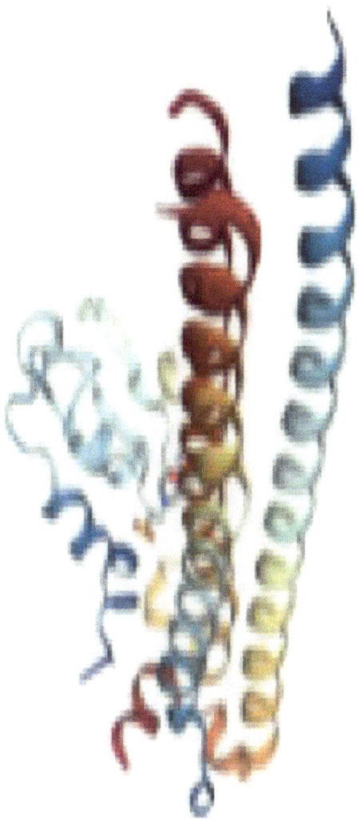

Fig. (10). Survivin.

Fig. (11). ACPA.

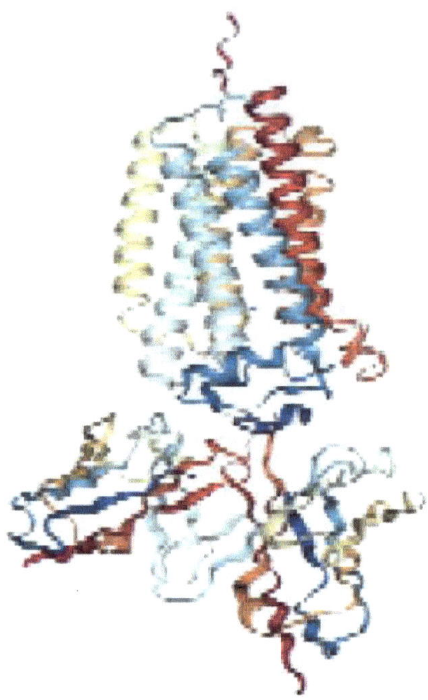

Fig. (12). Adiponectin.

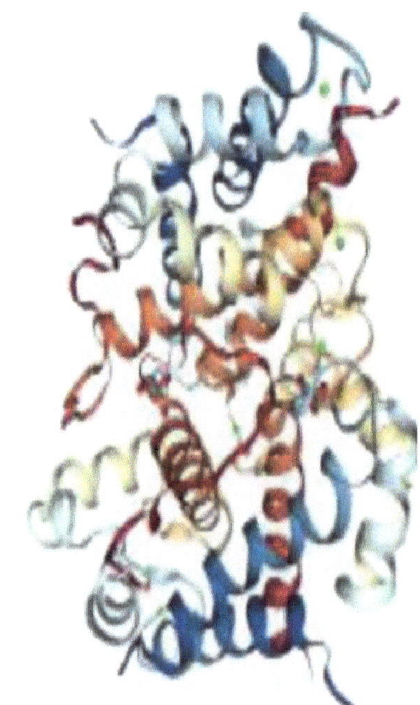

Fig. (13). Calprotectin.

The biomolecular analysis of the intermolecular interactions includes hydrophobic and hydrogen-bonding interactions. The trajectories of cluster analysis by Ligplot resulted in MK-3207 forming two hydrogen bonds with the target protein. Aside from hydrogen bonding interactions, the Protein was also involved in hydrophobic interactions with the ligand on the Protein's active site (Glu196, Lys11, Thr13, Gly10, Val22, Asp103, and Leu7).

CONCLUSION

Water pollution is a significant concern worldwide, as biological or chemical effects of water change with pollutants. There is an urgent need to investigate water pollution management techniques. It comprises three types: contamination by organic chemicals, inorganic substances *(e.g.,* heavy metals), and microbes. While some creatures are beneficial, others are detrimental and cause illness. Microorganisms can also limit the amount of oxygen in the water, making it less suitable for marine animal and plant life.

REFERENCES

[1]   R.P. Schwarzenbach, T. Egil, T.B. Hofstetter, U.V. Gunten, and B. Wehrli, "Global Water Pollution and Human Health", *Annu. Rev. Environ. Resour.,* vol. 35, pp. 109-136, 2010.

[2]   T. Dagan, M. Roettger, K. Stucken, G. Landan, R. Koch, P. Major, S.B. Gould, V.V. Goremykin, R. Rippka, N. Tandeau de Marsac, M. Gugger, P.J. Lockhart, J.F. Allen, I. Brune, I. Maus, A. Pühler, and W.F. Martin, "Genomes of Stigonematalean cyanobacteria (subsection V) and the evolution of oxygenic photosynthesis from prokaryotes to plastids", *Genome Biol. Evol.,* vol. 5, no. 1, pp. 31-44, 2013.
[http://dx.doi.org/10.1093/gbe/evs117] [PMID: 23221676]

[3]   S.F. Baldia, M.C.G. Conaco, T. Nishijima, S. Imanishi, and K.I. Harada, "Microcystin production during algal bloom occurrence in Laguna de Bay, the Philippines", *Fish. Sci.,* vol. 69, no. 1, pp. 110-116, 2003.
[http://dx.doi.org/10.1046/j.1444-2906.2003.00594.x]

[4]   R. Sinha, L.A. Pearson, T.W. Davis, M.A. Burford, P.T. Orr, and B.A. Neilan, "Increased incidence of Cylindrospermopsis raciborskii in temperate zones – Is climate change responsible?", *Water Res.,* vol. 46, no. 5, pp. 1408-1419, 2012.
[http://dx.doi.org/10.1016/j.watres.2011.12.019] [PMID: 22284981]

[5]   M.J. Harke, M.M. Steffen, C.J. Gobler, T.G. Otten, S.W. Wilhelm, S.A. Wood, and H.W. Paerl, "A review of the global ecology, genomics, and biogeography of the toxic cyanobacterium, Microcystis spp", *Harmful Algae,* vol. 54, pp. 4-20, 2016.
[http://dx.doi.org/10.1016/j.hal.2015.12.007] [PMID: 28073480]

[6]   A. Harishchander, and D.A. Anand, "Computational approach for identifying therapeutic micro RNAs", *Int. J. Pharm. Pharm. Sci.,* vol. 6, pp. 638-640, 2014.

[7]   G.M. Morris, R. Huey, W. Lindstrom, M.F. Sanner, R.K. Belew, D.S. Goodsell, and A.J. Olson, "AutoDock4 and AutoDockTools4: Automated docking with selective receptor flexibility", *J. Comput. Chem.,* vol. 30, no. 16, pp. 2785-2791, 2009.
[http://dx.doi.org/10.1002/jcc.21256] [PMID: 19399780]

[8]   O. Trott, and A.J. Olson, "AutoDock Vina: improving the speed and accuracy of docking with a new scoring function, efficient optimization, and multithreading", *J. Comput. Chem.,* vol. 31, no. 2, pp.

455-461, 2010.
[PMID: 19499576]

[9]   N.K. Sharma, and K.K. Jha, "Molecular docking: An overview", *Int. J. Adv. Sci. Res.,* vol. •••, pp. 67-72, 2010.

[10]   A.C. Wallace, R.A. Laskowski, and J.M. Thornton, "LIGPLOT: A program to generate schematic diagrams of protein-ligand interactions", *Protein Engin.,* vol. 8, pp. 127-134, 1995.

CHAPTER 8

Economic Consideration of an Off-Grid Hybrid Power Generation System using Renewable Energy Technologies: Case Study of an Institutional Area in the State of Rajasthan

Devendra Kumar Doda[1,*]

[1] *Department of Electrical Engineering, Vivekananda Global University, Jaipur, Rajasthan-303012, India*

Abstract: The focal point of this study is to recreate and plan a hybrid system consisting of a solar photovoltaic, a battery and a diesel generator and optimize the configuration into an off-grid hybrid structure to meet the electricity demand of an institutional area situated in Jaipur, Rajasthan, India. Various configurations have different specifications obtained to meet the load demand based on input parameters which are obtained from a pilot survey and the main survey a particular location. Various costing parameters such as per unit of cost and net present cost are estimated with the condition of meeting the maximum load demand. The HOMER (Hybrid Optimization Model for Electric Renewable) software is used for different simulation processes and finally it is found that solar PV-battery-diesel generator hybrid system is an economical system to meet the electricity demand in which the cost of energy is obtained as ₹ 13.83 and Net Present Cost is ₹ 9.78M with initial capital and operating costs of ₹ 4.20M and ₹ 646,319 per year, respectively. The diesel fuel cost is obtained as ₹ 5,09,288 per year. Meanwhile, the electricity produced and consumption are also estimated as 1,09,040 kWh/year and 81,939 kWh/year, respectively, with an unmet load of 1.77% only.

Keywords: Renewable energy resources, Hybrid power generation system, Optimization, HOMER, Cost of energy, Net Present Cost.

INTRODUCTION

Today, electricity is the basic requirement for the development of any country. The need for electricity is increasing enormously with rapidly increasing population growth. To fullfill the present and future need for power, new strategies are needed to develop in the developing nations like India. Hybrid

* **Corresponding author Devendra Kumar Doda:** Department of Electrical Engineering, Vivekananda Global University, Jaipur, Rajasthan-303012, India; E-mail:devendra.doda@gmail.com

systems with sustainable power source assets give an affordable and solid inventory of power bringing down per unit cost of energy delivered. Nowadays renewable energy sources like biomass, biogas, wind, solar photovoltaic, and micro-hydro are widely used in standalone as well as hybrid power generating systems of various combinations with a diesel generator as per the load demand and the availability of resources [1 - 9]. India's per capita power consumption has consistently been growing over time, from 734 kWh in 2008–09, the consumption has reached 1075 kWh in 2016, and finally, it has reached 1181 kWh in 2019, a prosperous growth of 60% in eleven years, *i.e.,* an average of 5.5% every year [10 - 16]. The conventional method is fulfilling the load requirements and peak load demands but it has some problems and constraints such as pollution, air contamination and global warming, high maintenance and operating cost of expensive equipment, lack of highly skilled labour; risk of severe accidents like boiler explosion, transmission line hazardous accidents, *etc.* So, in order to eliminate such problems, the concept of decentralized power generation with renewable energy sources is being utilised. Renewable energy includes solar, wind, biomass, geothermal, hydropower, and tidal energy forms, which give energy such as electricity generation, air, and water heating/cooling, and transportation services [17 - 28]. The hybrid combinations may be solar-wind, solar-biomass, wind-biomass, *etc.* for different locations based on the feasibility of various renewable energy resources. Sometimes diesel generators are also used with hybrid combinations to increase the overall reliability of the hybrid renewable energy system [29 - 36].

Table **1** represents the total power generated across the world per year from 2005 to 2019. It shows that the demand for electricity has continuously increased every year from 18,333TWh in 2005 to 25,721 TWh in 2019, hence there is a great need for electricity for global development. Tables **2** and **3** present the contribution of renewable energy resources and nation-level energy, respectively [10, 11].

Table 1. Global Electricity Generation (TWh) [10, 11].

Year	Electricity Generation (TWh)	Year	Electricity Generation (TWh)
2005	18,333	2013	23,375
2006	19,040	2014	23,765
2007	19,874	2015	23,950
2008	20,263	2016	25,082
2009	20,204	2017	25,570
2010	21,515	2018	25,648
2011	22,212	2019	25,721

(Table 1) cont.....

Year	Electricity Generation (TWh)	Year	Electricity Generation (TWh)
2012	22,709	-	-

Table 2. Contribution of Renewable Sources in Global Electricity Generation [10, 11].

Year	Electricity Generation (TWh)	Year	Electricity Generation (TWh)
2005	3278.93	2013	5041.65
2006	3437.04	2014	5295.24
2007	3552.73	2015	5513.01
2008	3804.49	2016	5859.62
2009	3886.53	2017	6231.93
2010	4187.10	2018	6673.49
2011	4398.93	2019	6943.84
2012	4722.20	-	-

Table 3. Nation Electricity Generation (TWh) [14].

Year	Electricity Generation (TWh)	Year	Electricity Generation (TWh)
2005	766.848	2013	1177.810
2006	768.492	2014	1271.872
2007	818.876	2015	1351.970
2008	840.048	2016	1433.392
2009	892.458	2017	1486.493
2010	976.432	2018	1546.517
2011	1056.838	2019	1588.261
2012	1056.838	-	-

Fig. **(1)** represents the total renewable contribution which is about 35.4%, comprising 10.1% from wind, 12.6% from large hydro, 8.7% from solar, 2.7% from biomass and 1.3% from small hydro and waste to power is about 0.04% together with the various non-renewable contributions such as coal 53.9%, gas 6.9%, nuclear 1.9%, diesel 0.2%.

RELATED LITERATURE REVIEW

There are different studies that are available in the literature mentioned by many researchers focused on single and/or hybrid renewable energy based on different locations and climate conditions. Some of them are presented in this section.

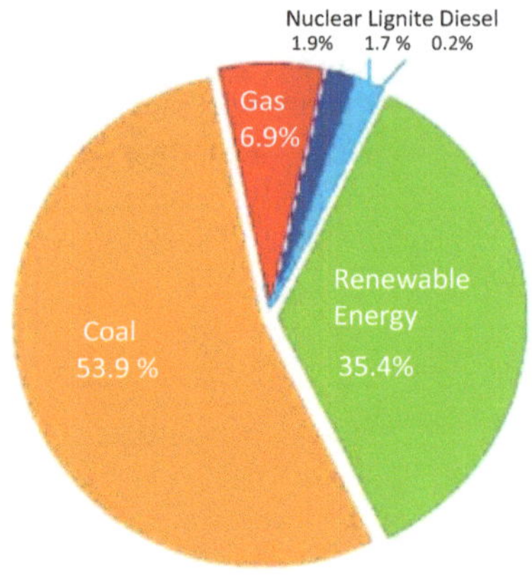
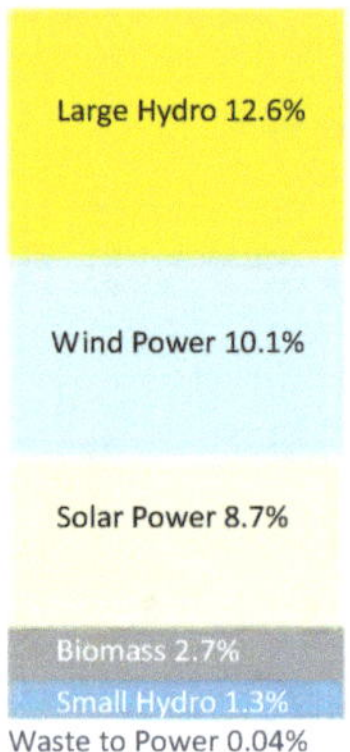

Fig. (1). Energy Scenarios in India [10 - 16].

The rooftop PV-battery system for the education campus at the Department of Electrical Engineering, NitteMeenaskshi Institute of Technology, Bengaluru, India, has been designed and simulated using HOMER software and ; estimated net present cost as ₹ 37100000 and the cost of energy as ₹ 14.8 for average energy consumption of 430 kWh/day [37]. Three configurations such as diesel ; diesel-PV; and diesel-PV-battery are considered for rural areas and observed that the diesel-PV-battery hybrid system is the optimum solution among all three systems. The cost of energy of the system is estimated as $ 0.377/kWh and reduces almost 720 tons/year of carbon dioxide. This system increases reliability but some mean extra costs due to the cost of diesel and fewer facilities available in rural areas [38]. A residential building at Suresh Gyan Vihar University, Jaipur, Rajasthan, India, is considered to have a total average load of 284 kWh/day. The total initial cost and operating cost are observed to be $181,983 and $4.776b per year, respectively, for PV-diesel hybrid systems having 160 kW PV and 15 kW diesel generators. The cost of the electricity produced is estimated to be $0.184/kWh [39]. At Kutubdia Island, Bangladesh, by using HOMER and RETscreen software, two renewable energy-based power generation systems are proposed. One is a solar PV-diesel energy system with 62kW PV arrays and a 9kW diesel generator and another wind-diesel energy system comprised of 51kW of wind turbines and 9kW of diesel generator. From the first system, the net annual GHG emission reduction is about 54.3t CO_2 and from the other system, it is about 42.9t CO_2. The first system is of lower payback and the second system is costlier. Using HOMER, the COE from solar PV-diesel is estimated to be $0.353/kWh and from wind-diesel, it is $0.487/kWh. The Wind-Diesel energy system is found to be the costlier option in terms of an equity payback period of about 11.2 years [26]. Optimization and economic evaluation of small-scale hybrid solar-wind power for remote areas in Egypt at the Electrical Engineering Department, Al Azhar University, Cairo, Egypt were evaluated and compared with stand-alone PV system using HOMER software. It is observed that the hybrid combination is more feasible and economical than the standalone PV system to supply electric power to remote areas. In the hybrid system, the number of batteries used is less and the net cost is less than the standalone PV system [7]. The cost of energy is estimated to be 6.24Rs/kWh using a wind-solar PV-grid and 21.81 Rs/kWh using a wind-solar PV-battery at a residential block in Maharashtra state, India using HOMER [6]. NPC and COE are estimated as Rs 1,82,77,696 and Rs.5.952, respectively in the off-grid system and saving in electricity bill using on-grid is estimated of INR 152136.845/month with INR1.211/unit. This system becomes more economic and reliable with diesel generators [40].

In this study, an off-grid PV-battery-diesel generator hybrid energy generation system is designed to fulfill the electricity demand and various output cost parameters are estimated for an institutional area situated in the state of Rajasthan, India (latitude and longitude coordinates are 26°45.9'N and 75°51.2'E respectively). The system is designed, analysed, and optimized with detailed resource analysis of the site using HOMER software.

OBJECTIVES AND INPUT DATAS REQUIRED

Objectives

The objectives of this study are as follows:

• To meet the required electricity demand of the selected locations using hybrid renewable energy-based electricity generation options.

• To design a hybrid system and optimise the system using HOMER software.

• To analyse the techno-economics from various output parameters obtained.

Input Data Required

The input data required for fulfilling the above objectives are as follows:

Resources Feasibility

Average Solar Radiation

In a year, Rajasthan has 300-330 days of clear sunshine which makes it suitable for power generation from solar radiation. Table **4** shows the average monthly solar radiation at the location [16]. Fig. (**2**) shows the graphical representation of the same.

Table 4. Average monthly solar radiation at a specific location.

Month	Solar radiation(kWh/m²/day)
January	4.177
February	4.929
March	5.750
April	6.437
May	6.834
June	6.902
July	5.680
August	5.395
September	5.870
October	5.352
November	4.423
December	3.897
Average monthly solar Radiation	**5.47**

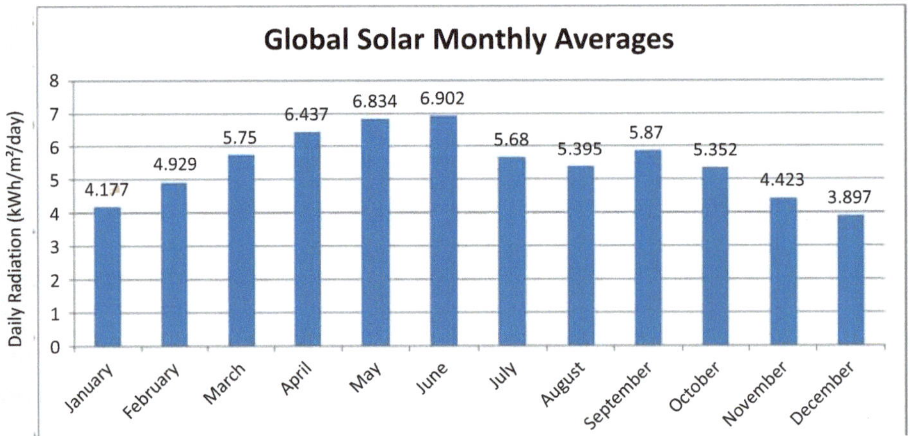

Fig. (2). Average monthly solar radiation at the selected location.

Average Wind Speed

Table **5** shows the monthly average wind speed [16].

Table 5. Monthly average wind speed at a specific location.

Months	Wind Speed
January	2.660
February	3.180
March	2.830
April	3.840
May	4.470
June	4.620
July	3.980
August	3.290
September	3.190
October	2.470
November	2.400
December	2.530
Annual Average	**3.29**

The average speed of wind is 3.29, which is below the standard value and not used to design the hybrid system.

Load Data

Table **6** represents the load data of the institute in both summer season (considered from the month of March to October) and winter seasons (Considered as the months of January, February, November, and December) which were collected from the pilot and main survey conducted at the location. The total load and peak load in summer season were obtained to be 277.24 kW and 46.67 kW, respectively. Similarly, the total load and peak load calculated in the winter season are 121.16 kW and 19.145 kW. The average load is estimated as 228.53 kWh per day. The load in the month of June is not considered in this study due to summer vacations. The graphical representation is also shown in Fig. (**3**).

Table 6. Load data for different months of specific locations.

Time (Hours)	Load (kW) in Winter Season	Load (kW) in Summer Season
00:00 - 01:00	0.000	0.000
01:00 – 2:00	0.000	0.000
02:00 - 03:00	0.000	0.000
03:00 - 04:00	0.000	0.000

(Table 4) cont.....

Time (Hours)	Load (kW) in Winter Season	Load (kW) in Summer Season
04:00 - 05:00	0.000	0.000
05:00 - 06:00	0.000	0.000
06:00 - 07:00	0.000	0.000
07:00 – 08:00	0.000	0.000
08:00 – 09:00	19.145	46.675
09:00 – 10:00	19.145	46.675
10:00 – 11:00	19.145	46.675
11:00 – 12:00	19.145	46.675
12:00 – 13:00	19.145	46.675
13:00 – 14:00	19.145	46.675
14:00 – 15:00	19.145	46.675
15:00 – 16:00	19.145	46.675
16:00 – 17:00	19.145	46.675
17:00 – 18:00	0.000	0.000
18:00 – 19:00	0.000	0.000
19:00 – 20:00	0.000	0.000
20:00 – 21:00	0.000	0.000
21:00 – 22:00	0.000	0.000
22:00 – 23:00	0.000	0.000
23:00 – 24:00	0.000	0.000

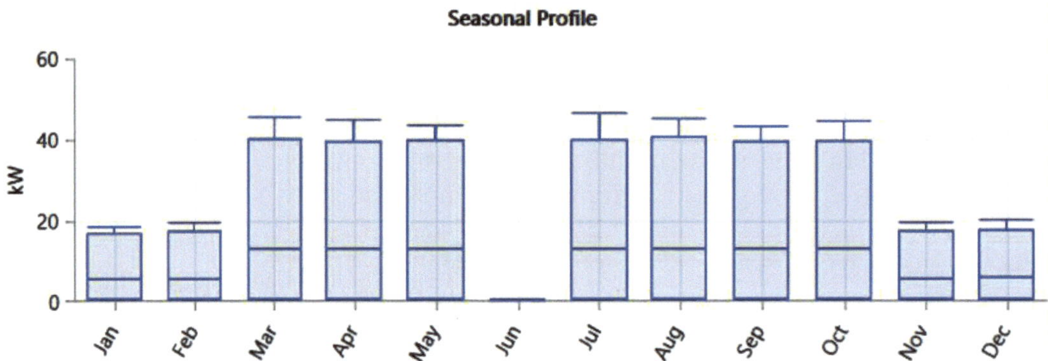

Fig. (3). Annual load profile of the selected locations.

ARCHITECTURE OF HYBRID SYSTEM

The simulation architecture of the solar PV-diesel generator hybrid system is presented in Fig. (**4**) using HOMER software. This system is designed to meet the entire load demand with minimum per unit cost of electricity and net present cost as well.

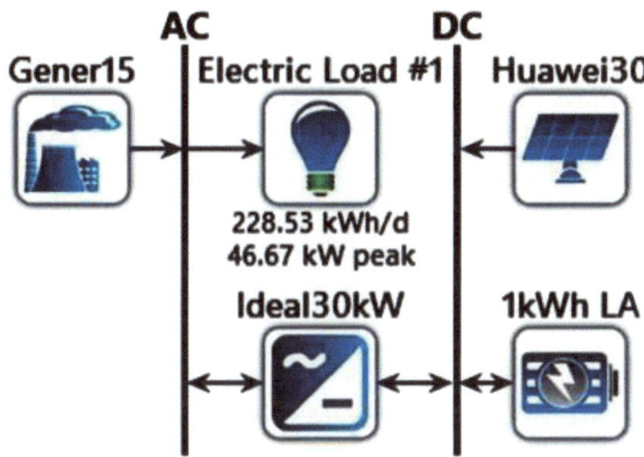

Fig. (4). Solar PV-battery-diesel generator hybrid system in HOMER.

Five different sizes of PV panels are considered with a battery bank having 73 strings each of nominal capacity 83.4 Ah and nominal voltage 12 V. The specification and costing parameters of PV panels and battery are shown in Tables **7** - **9**, respectively.A diesel generator (DG) of 15 kW size is also connected. The specification and costing parameter of DG are shown in Table **10**. Four converters of different sizes (1, 10, 20, and 30 kW) are considered for optimization. The specification and costing parameter of the converters are shown in Table **11**.

Table 7. Costing Parameters of PV panel [41].

Description	Specifications			
Size(kW)	1	10	20	25
Capital Cost (₹)	50000.00	530000.00	940000.00	1175000.00
Replacement cost (₹)	25000.00	400000.00	738000.00	900000.00
O & M cost (₹)	1000.00	10600.00	18800.00	23500.00
Life Time	20 years			
Derating Factor	96.00			
Efficiency	17.30%			
Ground reflectance	20%			

Table 8. General specifications of the battery [42].

Description	Specifications
Name	Generic 1kWh Lead Acid
Manufacturer	Generic
Nominal Capacity(Ah)	83.4
Nominal Voltage(V)	12
Round trip efficiency (%)	80
Minimum state of charge (%)	40
Float Life (Years)	4.00
Maximum charge rate (A/Ah)	1
Maximum charge current (A)	16.7
Lifetime throughput (kWh)	800.00

Table 9. Costing parameters of the battery [43].

Description	Specifications
Quantity(strings)	73
Capital Cost (₹)	10000.00
Replacement cost (₹)	500.00
O & M cost (₹)	200.00
Life Time (Years)	4.00

Table 10. Costing parameters of diesel generator [43, 44].

Description	Specifications
Size (kW)	15
Capital Cost (₹)	150000
Replacement cost(₹)	150000
O & M cost (₹)	15.000
Life Time (Hours)	15000

Table 11. Specifications and costing parameters of converters [44].

Description	Specifications			
Size(kW)	1	10	20	30
Capital Cost (₹)	50000.00	250000.00	350000.00	500000.00
Replacement cost (₹)	35000.00	200000.00	300000.00	450000.00
O & M cost (₹)	1000.00	5000.00	7000.00	10000.00

(Table 11) cont.....

Description	Specifications
Life Time (Years)	20
Efficiency (%)	96.00

Specification of Fuel Used

Fuel used: Diesel

Cost (Rupees per litre): 70.00

Intercept Coefficient (L/hr/kW rated): 0.635

Slope (L/hr/kW output): 0.327

Lower heating value (MJ/kg): 43.2

Density (kg/m3): 820

Carbon content (%): 88

Sulfur content (%): 0.4

RESULTS AND DISCUSSION

An off-grid hybrid system of 75 kW solar PV-battery-diesel generator is designed and simulated for an institutional area situated in Jaipur, Rajasthan. Various combinations of system specifications are considered for getting the optimized results. In this section, various optimization results based on input data and different simulations in HOMER of solar PV-battery-diesel generator hybrid system are presented and discussed is shown in Fig. (**5**). The minimum cost of electricity is obtained as ₹ 13.83 and the net present cost as ₹ 9.78M with an initial capital cost of ₹ 4.20M and operating cost of ₹ 646,319/year. The annual diesel fuel cost is obtained as ₹ 5,09,288 is shown in Fig. (**6**). The annual electricity production and consumption are estimated as ₹ 1,09,040 kWh and ₹ 81,939 kWh with unmet load 1.77% only. The total pollutant emitted by the hybrid system is calculated as 19365.66 kg/year which consists of carbon dioxide 19030 kg/year, carbon monoxide 129 kg/year, unburned hydrocarbons 5.24 kg/year, particulate matter 7.82 kg/year, sulfur dioxide 46.6 kg/year, and nitrogen oxides 147 kg/year.

Sensitivity		Architecture							Cost					
Project Lifetime (years)	Capacity Shortage (%)	Huawei30 (kW)	Huawei30-MPPT (kW)	Gener15 (kW)	1kWh LA	Ideal30KW (kW)	Dispatch	COE (₹)	NPC (₹)	Operating cost (₹/yr)	Initial capital (₹)	Fuel cost (₹/yr)	O&M (₹/yr)
20.0	10.0	60.0	30.0	15.0	73	30.0	CC	₹ 13.83	₹ 9.78M	₹ 6,46,319	₹ 4.20M	₹ 5,09,288	₹ 1,15,170
22.0	10.0	60.0	30.0	15.0	73	30.0	CC	₹ 13.79	₹ 10.1M	₹ 6,58,015	₹ 4.20M	₹ 5,09,288	₹ 1,15,170
20.0	15.0	50.0	30.0	15.0	71	25.0	LF	₹ 13.64	₹ 9.52M	₹ 6,82,028	₹ 3.64M	₹ 5,63,067	₹ 99,835
22.0	15.0	50.0	30.0	15.0	71	25.0	LF	₹ 13.60	₹ 9.79M	₹ 6,91,684	₹ 3.64M	₹ 5,63,067	₹ 99,835
20.0	20.0	60.0	30.0	15.0	67	30.0	LF	₹ 12.71	₹ 8.92M	₹ 5,53,569	₹ 4.14M	₹ 4,25,199	₹ 1,09,755
22.0	20.0	60.0	30.0	15.0	67	30.0	LF	₹ 12.67	₹ 9.17M	₹ 5,64,975	₹ 4.14M	₹ 4,25,199	₹ 1,09,755
20.0	5.00	60.0	30.0	15.0	101	30.0	CC	₹ 14.21	₹ 10.1M	₹ 6,48,964	₹ 4.48M	₹ 5,05,921	₹ 1,19,330
22.0	5.00	60.0	30.0	15.0	101	30.0	CC	₹ 14.16	₹ 10.4M	₹ 6,60,628	₹ 4.48M	₹ 5,05,921	₹ 1,19,330

Fig. (5). Sensitivity variables and related results obtained in HOMER.

Architecture						Cost							System		
Huawei30 (kW)	Huawei30-MPPT (kW)	Gener15 (kW)	1kWh LA	Ideal30kW (kW)	Dispatch	COE (₹)	NPC (₹)	Operating cost (₹/yr)	Initial capital (₹)	Fuel cost (₹/yr)	O&M (₹/yr)	Ren Frac (%)	Total Fuel (L/yr)	Cap Short (%)	Cap Short (kWh/yr)
60.0	30.0	15.0	73	30.0	CC	₹ 13.83	₹ 9.78M	₹ 6,46,319	₹ 4.20M	₹ 5,09,288	₹ 1,15,170	78.3	7,276	10.0	8,351
		15.0	423	25.0	CC	₹ 41.46	₹ 28.0M	₹ 2.69M	₹ 4.81M	₹ 2.41M	₹ 1,86,250	0	34,418	8.90	7,421

Fig. (6). Optimization results obtained in HOMER.

CONCLUSION AND FURTHER RECOMMENDATIONS

Based on the pilot and main survey conducted, the electricity demand is calculated in two seasons summer months and winter months at a selected location situated in Jaipur district in the state of Rajasthan. The feasibility of renewable energy resources such as solar radiation and wind speed is analyzed based on monthly average data at the selected location. An off-grid hybrid system is designed with solar PV panels with different battery specifications with diesel generator to fulfill the overall load demand and peak load demand of institutional area with minimum cost of electricity and net present cost as well. The optimization results of the hybrid system consist of various cost parameters such as the cost of electricity (₹ 13.83), net present cost (₹ 9.78M), initial capital cost (₹ 4.20M), operating cost (₹ 646,319/year) and diesel fuel cost (₹ 5,09,288/year). Meanwhile, the electricity production (1,09,040 kWh/year) and consumption (₹ 81,939 kWh/year) were also assessed with unmet load of 1.77% only.

This system can be further extended by adding biomass and wind energy, thereby increasing its reliability and making it pollution free. However, it should be implemented to understand the realistic challenges and resolutions which will be helpful to planning engineers and policy makers for any other sites like rural areas.

REFERENCES

[1]   J. Chang, and S.Y. Jia, "Modeling and application of wind-solar energy hybrid power generation system based on multi-agent technology", *Proc. 2009 Int. Conf. Mach.Learn. Cybern,* vol. 3, pp. 1754-1758, 2009.

[2]   P. Anjana, and H.P. Tiwari, "Electrification by mini hybrid PV-solar/wind energy system for rural, remote and Hilly/trible areas in Rajasthan (India)", *DRPT 2011- 2011 4th Int. Conf. Electr. Util. Deregul. Restruct. Power Technol.,* pp. 1470-1473, 2011.

[3]   A. Shrivastava, D.K. Doda, and M. Bundele, "Economic and environmental impact analyses of hybrid generation system in respect to Rajasthan", *Environ. Sci. Pollut. Res. Int.,* vol. 28, no. 4, pp. 3906-3912, 2021.
[http://dx.doi.org/10.1007/s11356-020-10041-6] [PMID: 32656752]

[4]   S. Georges, and F.H. Slaoui, "Case study of hybrid wind-solar power systems for street lighting", *Proc. ICS Eng 2011 Int. Conf. Syst. Eng.,* pp. 82-85, 2011.
[http://dx.doi.org/10.1109/ICSEng.2011.22]

[5]   D.K. Doda, M. Bundele, A. Shrivastava, and T.C. Kandpal, "Financial feasibility of solar PV lanterns for households of a remote village cluster without access to electricity", *Int. J. Environ. Sustain. Dev.,* vol. 20, no. 3/4, pp. 354-365, 2021.
[http://dx.doi.org/10.1504/IJESD.2021.116865]

[6]   K.K. Jagtap, G. Patil, P.K. Katti, and S.B. Kulkarni, "Techno-economic modeling of wind-solar PV and wind-solar PV-biomass hybrid energy system", *IEEE Int. Conf. Power Electron. Drives Energy Syst. PEDES 2016,* pp 1-6,2017.

[7]   I. Elsayed, I. Nassar, and F. Mostafa, "Optimization and economic evaluation of small scale hybrid solar/wind power for remote areas in Egypt", *19th Int. Middle-East Power Syst. Conf. MEPCON 2017-Proc.,* vol. 2018, pp. 25-30, 2018.

[8]   K. Shekhawat, D.K. Doda, A.K. Gupta, and D.M. Bundele, "Decentralized Power Generation using Renewable Energy Resources: Scope, Relevance and Application", *Int. J. Innov. Technol. Explor. Eng.,* vol. 8, no. 9, pp. 3052-3060, 2019. https://www.ijitee.org/wp-content/uploads/papers/v8i9/I8595078919.pdf10.35940/ijitee.I8595.078919

[9]   H.M. Sharma, and D.K. Doda, "Scope of Decentralized Power Generation using Renewable Energy Resources at Global Level-A Review & Survey at a Glance", *Journal of Automation & Systems Engineering,* vol. 11, no. 4, pp. 280-294, 2017.

[10]   All India installed capacity of power stations. 2021. https://en.wikipedia.org/wiki/Electricity_sector_in_India

[11]   Central Electricity Authority-Growth of Electricity Sector in India. 2020. http;//growth_2020.pdfcea.nic.in

[12]   International Energy Agency. 2021. https://www.iea.org/countries/india

[13]   Statistical Review of World Energy. 2020. https://www.bp.com/en/global/corporate/energy-economics/statistical-review-of-world-energy.html

[14]   National Electricity Plan (Generation). https://policy.asiapacificenergy.org/node/3641

[15]   Rajasthan-Solar-Energy-Policy-2014. https://mnre.gov.in/filemanager/UserFiles/Grid-Connecte--Solar-Rooftop-policy/Rajasthan-Solar-Energy-Policy-2014.pdf

[16]   NASA Prediction of Worldwide Energy Resources. 2020. https://power.larc.nasa.gov/downloads/POWER_SinglePoint_Daily_20170501_20180430_026d77N_075d84E_dc2f8fc0.txt

[17]   N. Gothwal, T. Manglani, and D. Kumar, "Importance of Off-Grid Power Generation using Renewable Energy Resources - A Review", *Int. J. Comput. Appl.,* vol. 179, no. 28, pp. 38-41, 2018.
[http://dx.doi.org/10.5120/ijca2018916634]

[18] A. Mekonnen, "Optimal Sizing of Grid Connected PV/Wind Hybrid System Using Homer Software", *Int. J. Sci. Res.,* vol. 8, no. 1, pp. 482-488, 2019.

[19] K.K. Jagtap, G. Patil, P.K. Katti, N.N. Shinde, and V.S. Biradar, "Techno-economic feasibility study of wind- solar PV hybrid energy system in maharashtra state, India", *IEEE Int. Conf. Power Electron. Drives Energy Syst. PEDES 2016,* pp 1-5, 2017.

[20] S. Kumar, D.K. Doda, and D. Chauhan, "A comparative analysis of different thermal parameter arrangements with silicon polycrystalline using PVSYST", *Int. J. Eng.Appl. Sci.,* vol. 4, no. 7, pp. 20-29, 2017.

[21] S. Ghose, A. El Shahat, and R.J. Haddad, "Wind-solar hybrid power system cost analysis using HOMER for Statesboro, Georgia", *Conf.Proc. IEEE Southeastcon,* vol. 8, pp. 3-5, 2017.

[22] W. Astatike, and P. Chandrasekar, "Design and performance analysis of hybrid micro-grid power supply system using HOMER software for rural village in adama area, Ethiopia", *Int. J. Sci. Technol. Res.,* vol. 8, no. 6, pp. 267-275, 2019.

[23] K.E. Okedu, and R. Uhunmwangho, "Optimization of renewable energy efficiency using HOMER", *Int. J. Renew. Energy Res.,* vol. 4, no. 2, pp. 421-427, 2014.

[24] R. Ranjan, D.K. Doda, M. Lalwani, and M. Bundele, "Simulation and Optimization of Solar Photovoltaic-Wind- Diesel Generator standalone Hybrid system in Remote Village of Rajasthan, India", *International Conference on Artificial Intelligence: Advances and Applications: Algorithms for Intelligent Systems,* Springer,Singapore, pp 279-286, 2019.
[http://dx.doi.org/10.1007/978-981-15-1059-5_31]

[25] R. Srivastava, and V.K. Giri, "Optimization of hybrid renewable resources using HOMER", *Int. J. Renew. Energy Res.,* vol. 6, no. 1, pp. 157-163, 2016.

[26] S. Salehin, M.T. Ferdaous, R.M. Chowdhury, S.S. Shithi, M.S.R.B. Rofi, and M.A. Mohammed, "Assessment of renewable energy systems combining techno-economic optimization with energy scenario analysis", *Energy,* vol. 112, pp. 729-741, 2016.
[http://dx.doi.org/10.1016/j.energy.2016.06.110]

[27] C. Lao, and S. Chungpaibulpatana, "Techno-economic analysis of hybrid system for rural electrification in Cambodia", *Energy Procedia,* vol. 138, pp. 524-529, 2017.
[http://dx.doi.org/10.1016/j.egypro.2017.10.239]

[28] A. Oulis Rousis, D. Tzelepis, I. Konstantelos, C. Booth, and G. Strbac, "Design of a hybrid ac/dc microgrid using homer pro: Case study on an islanded residential application", *Inventions (Basel),* vol. 3, no. 3, p. 55, 2018.
[http://dx.doi.org/10.3390/inventions3030055]

[29] M. Mehrpooya, M. Mohammadi, and E. Ahmadi, "Techno-economic-environmental study of hybrid power supply system: A case study in Iran", *Sustain. Energy Technol. Assess.,* vol. 25, pp. 1-10, 2018.
[http://dx.doi.org/10.1016/j.seta.2017.10.007]

[30] S. Kumar, D.K. Doda, and D. Chauhan, "A Review Paper on Performance Analysis of Various HPVT / PVT Modules under Different Indian Climatic Conditions", *Int. J. Adv. Res. Comput. Sci. Softw. Eng.,* vol. 7, no. 6, pp. 488-498, 2017.
[http://dx.doi.org/10.23956/ijarcsse/V7I6/0298]

[31] R. Ranjan, B. Modi, and D.K. Doda, "Distributed Generation of Power using Renewable Energy Resources-A Comparative Review of Grid-connected & Stand-alone System", *Int. J. Eng. Sci. Res. Technol.,* vol. 5, no. 3, pp. 641-646, 2016.

[32] J. Gautam, M.I. Ahmed, and P. Kumar, "Optimization and comparative analysis of solar-biomass hybrid power generation system using homer", *2nd International Conference on Intelligent Circuits and Systems* pp 401-405, 2018.
[http://dx.doi.org/10.1109/ICICS.2018.00087]

[33]   R. Uhunmwangho, M. Odje, and K.E. Okedu, "Comparative analysis of mini hydro turbines for Bumaji Stream, Boki, Cross River State, Nigeria", *Sustain. Energy Technol. Assess., vol. 27, pp. 102-108, 2018.
[http://dx.doi.org/10.1016/j.seta.2018.04.003]

[34]   N. Shipta, and Hr. Sridevi, "Optimum design of Rooftop PV System for An Education Campus Using HOMER", *2019 Global Conference for Advancement in Technology (GCAT),* Bangalore, India,pp 1-4,2019.

[35]   N. Gothwal, T. Manglani, D.K. Doda, D.K. Somvanshi, and M. Bundele, "Design and Optimization of PV-Wind-DG and Grid based Hybrid System for an Educational Institute in India", *International Conference on Artificial Intelligence: Advances and Applications: Algorithms for Intelligent Systems* Springer,Singapore, pp 131-139, 2019.
[http://dx.doi.org/10.1007/978-981-15-1059-5_16]

[36]   H.M. Sharma, D.K. Doda, and M. Bundele, "Proposed an Optimize Off-Grid Hybrid Model using Solar Photovoltaic-Wind-DG Technologies for the Climate Conditions of the State of Rajasthan, India", *IEEE International Conference on Recent Advances and Innovations in Engineering,* 2018.
[http://dx.doi.org/10.1109/ICRAIE.2018.8710438]

[37]   N. Shiila, and Hr. Sridevi, "Optimum design of rooftop PV system for an education campus using HOMER", *Global Conference for Advancement in Technology (GCAT)* Bangalore, India, pp 1-4,2019.

[38]   C. Lao, and S. Chungpaibulpatana, "Techno-economic analysis of hybrid system for rural electrification in Cambodia", *Energy Procedia,* vol. 138, pp. 524-529, 2017.
[http://dx.doi.org/10.1016/j.egypro.2017.10.239]

[39]   M. Jain, and N. Tiwari, "Optimization and Simulation of Solar Photovoltaic cell using HOMER: A Case Study of a Residential Building", *Int. J. Sci. Res.,* vol. 3, pp. 1221-1223, 2014.

[40]   D. K. Doda, A. Shrivastava, and M. Bundele, "Design & optimization of a power generation system based on renewable energy technologies", *Int. J. Innov. Technol. Explor. Eng.,* vol. 8, no. 11, pp. 1134-1138, 2019.
[http://dx.doi.org/10.35940/ijitee.J1304.0981119]

[41]   The Sukum Solar Power Company., http://sukam-solar.com/wp-content/uploads/Su-Kam-Data-Sheet-for-Solar-Panels-Poly-Rev-02Nov2016.pdf

[42]   The Exide Industries-NDP Tubular Standby batteries., http://docs.exideindustries.com/pdf/ industrial-batteries/NDP-Tubular-Range.pdf

[43]   The Kirloslar Oil Engines., http://www.koelgreen.com/5-kva-125-kva

[44]   The Sukum Solar Power Company., http://www.su-kam.com/Upload/UpProductCatalogue/2017-2 COLOSSALSeriesCatalogue.pdf

CHAPTER 9

High Optimization of Image Transmission and Object Detection Technique for Wireless Multimedia Sensor Network

R. Kabilan[1,*], Ravi R.[2], J. Zahariya Gabrie[2] and M. Philip Austin[1]

[1] *Department of ECE, Francis Xavier Engineering College, Affiliated with Anna University, 103/G2, Bypass Road, Vannarpettai, Tirunelveli, Tamil Nadu 627003, India*

[2] *Department of Electronics and Communication Engineering, Francis Xavier Engineering College, Tirunelveli, India*

Abstract: One of the most important issues in Wireless Multimedia Sensor Networks is the energy efficiency of object detection and image transmission. In-node object detection and tracking algorithms have been proposed in recent WMSN approaches. However, with a little effort, the WMSN will be able to detect the presence and absence of objects in images. For the WMSN, a new approach for the above technique is suggested in this research. Instead of sending a whole image, this technique sends image parts. It ensures energy saving inside the node and minimum picture content which is transferred to the sink node. On the basis of in-node reconstructed and energy consumption picture, it suggests that the technique is evaluated using (PSNR). In comparison with existing state-of-the-art methodologies, simulation results demonstrate that the suggested methodology saves 95 percent of node energy with a received picture PSNR of 46 dB.

Keywords: WMSN, PSNR, SNR, transmission, the detection algorithm.

INTRODUCTION

Any sort of signal processing in which the input is an image, such as a photograph or a video frame, is known as image processing. The result of image processing might be a picture or a set of image-related features or parameters. The majority of image processing approaches consider the picture as a two-dimensional signal that is then processed using traditional signal processing techniques. Although digital image processing is the most common, optical and analogue image proces-

---

* **Corresponding author R. Kabilan:** Department of ECE, Francis Xavier Engineering College, Affiliated with Anna University, 103/G2, Bypass Road, Vannarpettai, Tirunelveli, Tamil Nadu 627003, India; E-mail: rkabilan13@gmail.com

sing are also feasible [1-6]. The output of a segmentation step, which is normally raw pixel data, is nearly always followed by a representation and description, forming either the region's boundary or all points within the region. Choosing a representation is simply one aspect of the process of translating raw data into a format that can be processed by a computer. Lossy or lossless image compression is possible. For archiving purposes, lossless compression is desirable, as it is for medical imaging, technical drawings, clip art, and comics. Compression artefacts are created by lossy compression algorithms, especially when utilised at low bit rates [7]. Lossy approaches are best for natural pictures like photographs, when a small (often undetectable) loss of quality is acceptable in exchange for a significant drop in the bit rate [8-15]. The fundamental objective of image compression is to get the highest picture quality at a given bit rate (or compression rate), however, there are other essential aspects of image compression schemes: Generally speaking, it refers to a loss in quality produced by manipulating the bit stream or file (without decompression and recompression). Scalability is also known as progressive coding or embedded bit stream. Scalability may also be seen in lossless codecs, commonly in the form of coarse-to-fine pixel scans, despite its contradictory nature. Scalability is particularly helpful for evaluating images while they are being downloaded (for example, in a web browser) or enabling varying quality access to databases. Histogram equalisation (HE) is a widely used technique for improving picture contrast. Its core concept is to map grey levels using the probability distribution of the input grey levels as a guide. It improves the overall contrast by flattening and stretching the dynamical range of the image's histogram. Medical image processing and radar image processing are two examples of where HE has been used.

The characteristics of interest in certain CT radiographs only occupy a small portion of the grey scale. Contrast enhancement is a technique for increasing the contrast of certain elements so that they occupy a larger amount of the displayed grey level range while preserving the overall image quality. Contrast enhancement approaches aim to find the best transformation function between the original grey level and the projected intensity so that the contrast between neighboring structures in a picture is maximized [16-20].

Three-dimensional facial recognition is a new trend that claims to attain higher accuracy. This method captures information on the geometry of a face using 3D sensors. The shape of the eye sockets, nose, and chin are among the characteristic features on the surface of a face that are identified using this information.

One advantage of 3D facial recognition is that, unlike other systems, it is unaffected by changes in illumination. It can also recognise a face from a variety of angles, even from a profile perspective. The precision of facial recognition is much improved when three-dimensional data points from a face are used. The development of advanced sensors that capture 3D facial imagery better is enhancing 3D research. Structured light is projected onto the face by the sensors. On a single CMOS chip, up to a dozen or more of these image sensors may be put, each capturing a distinct part of the spectrum [21-26].

RELATED WORK

Target Tracking by Particle Filtering in Binary Sensor Networks

One of the most intriguing uses for wireless sensor networks is surveillance and other key areas such as embedded computing, computer vision, and image sensors. As a result, research in areas such as video processing algorithms, distributed engines, and power management, is still needed for wireless surveillance systems. We present a wireless smart camera with an infrared sensor and a solar energy harvester in this research.

Image Texture Feature Extraction Using GLCM Approach

Structure functions are generated in a statistical structure study from the mathematical submission of detected mixes of extremes at defined roles in comparison to each other in the image. Research is classified as first-order, second-order, or higher-order based on the number of strength components (pixels) in each combination. The Grayish Level Co-occurrence Matrix (GLCM) approach is a technique for obtaining mathematical structural functions in the second order. The method has been implemented by a number of different programmers. The matrix element P I j (x, y) represents the relative frequency with which two pixels are separated by a pixel distance (x, y), one with intensity I and the other with intensity 'j', appear within a particular neighborhood. The second-order statistical probability values for changes between grey levels I and 'j' at a specific displacement distance d and at a particular angle () are included in the matrix element P I j | d,). Using a high number of intensity levels, G necessitates keeping a lot of temporary data, *i.e.* a G G matrix for each (x, y) or (d, d) combination.

Co-Occurrence Matrix and its Statistical Features

From n inputs, a k-winner-take-all (kWTA) network can discover the k greatest numbers. To implement the kWTA method, a dual neural network (DNN) solution has been presented. The DNN technique has a substantially lower number of links than the traditional way. The convergence time of the DNN-kWTA model, stated in terms of input variables, was given a rough upper bound.

The actual convergence time of the DNN-kWTA model is calculated in this short. With these findings, we can investigate the convergence time without having to spend too much time simulating the network dynamics. The statistical features of the convergence time when the inputs are evenly distributed are also studied theoretically. Because a nonuniform distribution may be changed to a uniform one while preserving the inputs' ordering, our theoretical solution holds for nonuniformly distributed inputs as well.

PROPOSED SYSTEM

We can now combine diverse sensing devices into a single sensing platform thanks to advancements in miniaturized technology. The availability of CMOS cameras, which convert WSN into Wireless Multimedia Sensor Networks, is a significant benefit of this downsizing (WMSN). This transformation enables multidimensional signal processing methods to be implemented on these sensor systems. As a result, it provides much more services than a typical WSN. Surveillance, habitat monitoring, traffic control, intrusion detection, and healthcare monitoring are all possible applications. They can also be employed in tough, unsupervised, or under-resourced places. The work in this study focuses on changing the existing paradigm of data processing in WMSN to achieve this coherency. The goal is to reduce the in-node processing cost in order to extend the node lifespan.

Kalman Filter

A Kalman filter is an optimum estimator, meaning it infers parameters of relevance from erroneous, imprecise, and uncertain data. It is recursive, which means it can handle new metrics as they come in. (Comparing batch processing, which requires all data to be present.)

Kalman Filter Extensions

• Outlier measurements are rejected by validation gates.

• Independent measurement processing is serialized to avoid asymmetric covariance matrices.

• Numerical rounding difficulties are avoided by serialization of independent measurement processing.

• Lines appearing in the Kalman filter due to nonlinear problems.

Processing of Sequential Measurements: If the noise vector components of the measurement are uncorrelated, state updating can be done by one measurement at a time. As a result, scalar inversions take the role of matrix inversions. Procedure: scalar measurements are processed consecutively (in any order) using scalar measurement equations, just as they were processed previously.

Extended Kalman Filter (EKF)

Non-linear state updating or measurement equations are found in many practical systems. With a loss of optimality, the Kalman filter can be applied to a linearized version of these equations.

Iterated Extended Kalman Filter (IEKF)

As an operational point, the EKF linearizes the state and measurement equations around the expected state. In actuality, this forecast is frequently incorrect. Re-evaluating the filter around the new estimated state functioning point can improve the estimate. The IEKF is a refining method that can be repeated until a little more improvement is gained.

Flow Chart of Robust Detector (Fig. 1)

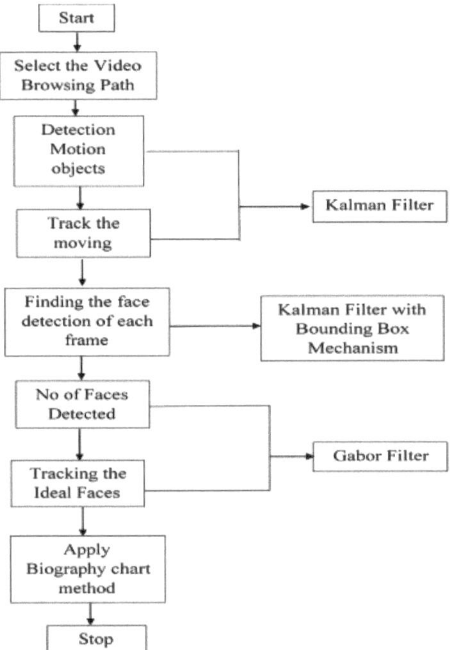

Fig. (1). Robust detector flow chart.

Gabor Filter

A Gabor filter, named after Dennis Gabor, is a linear filter used for edge detection in image processing. Gabor filters' frequency and orientation representations are comparable to those of the human visual system, and they've been shown to be very useful for texture representation and discrimination. A 2D Gabor filter is a Gaussian kernel function modulated by a sinusoidal plane wave in the spatial domain.

REQUIREMENT

A. HARDWARE REQUIREMENT

CPU type: Intel Pentium 4

Clock speed: 3.0 GHz

RAM: 512 MB

Hard disk capacity: 80 GB

Monitor type: 15 Inch color monitor

Input Device: Standard Keyboard and Mouse

Output Device: VGA and High Resolution Monitor

Compact Disk: 650 MB

B. SOFTWARE REQUIREMENT

Operating system: Windows 7

Coding Language: MATLAB 2014A

Design Software: Rational Rose Enterprise Edition

Documentation: Microsoft Office

SIMULATION RESULTS

Input File

The input video file is shown in Fig. (2). The video file in the input is in the.avi format. Here, the news2.avi video is used as the input for recognizing the faces in the video. Then, using a strong object appearance detector approach, faces are recognized from the input video frames.

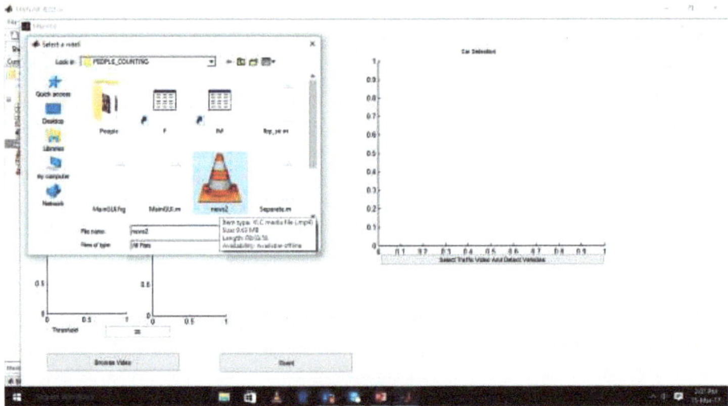

Fig. (2). Input file.

Face (Object) Detection

The face (object) detection of the input video is shown in Fig. (3). The news reader's face is detected and presented in a yellow colour box in this video of a news reader.

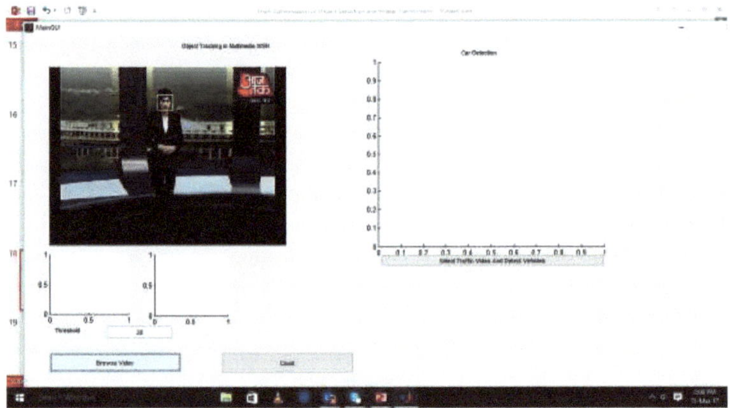

Fig. (3). Face (Object) Detection.

Total Number of Faces

The total number of faces discovered is shown in Fig. (**4**). There are a total of 26 faces spotted in the News 2 video utilizing a robust object appearance detector using a Kalman filter technique.

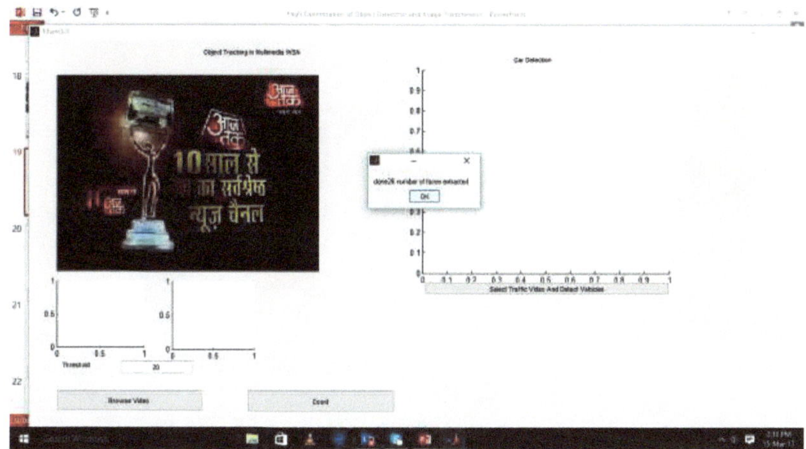

Fig. (4). Total number of faces.

Object Matching with Training Set

Objects matching the training set are shown in Fig. (**5**). The item that is face matched with the training set in this case is 6.

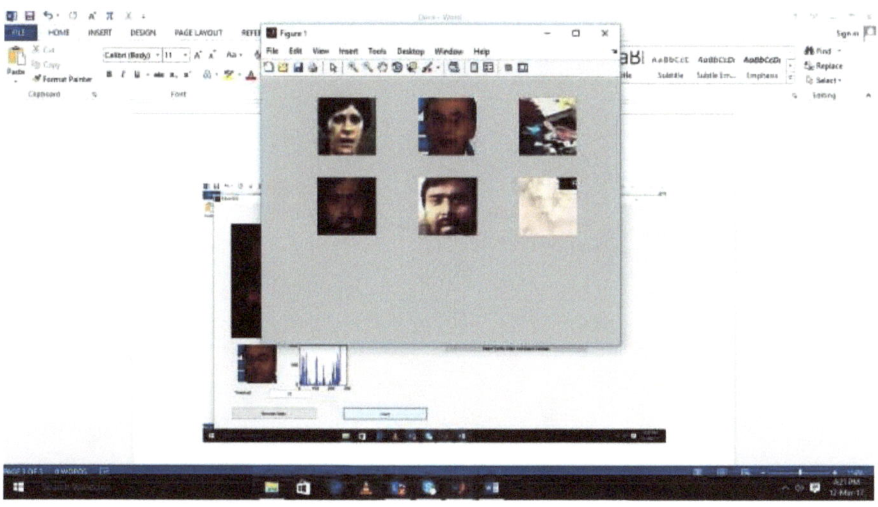

Fig. (5). Object matching with training set.

Histogram for Detected Faces

The histogram for the identified faces is shown in Fig. (**6**). The pixel value is represented graphically by the histogram. The threshold value is 20 in this case.

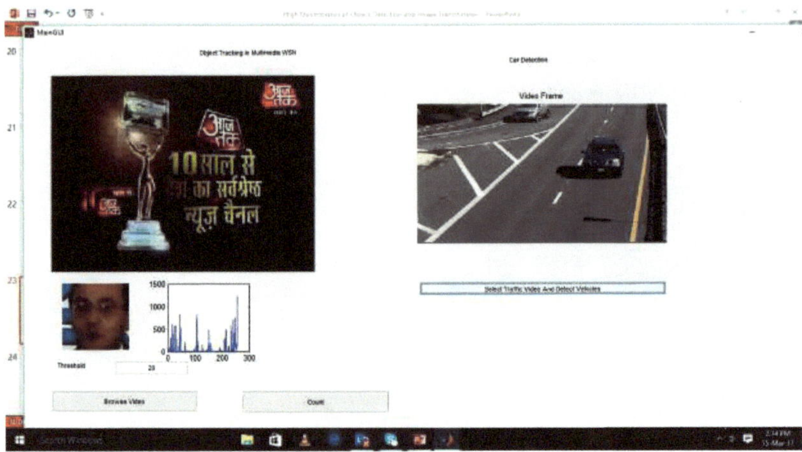

Fig. (6). Histogram for detected faces.

Biograph Model

Fig. (**7**) shows a biograph model. Here the detected faces are transmitted *via* nodes using a spanning tree based approach.

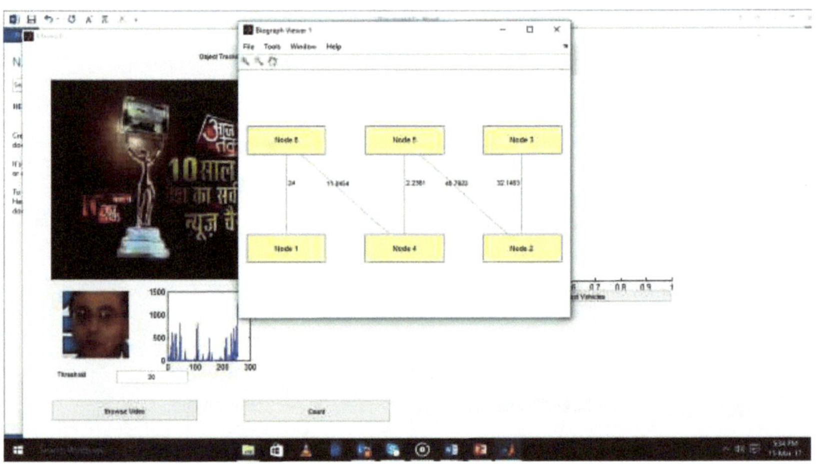

Fig. (7). Biograph model.

Input Video 2

Input video 2 is shown in Fig. (**8**). The input is a video of an automobile racing race. The input is in the avi file format.

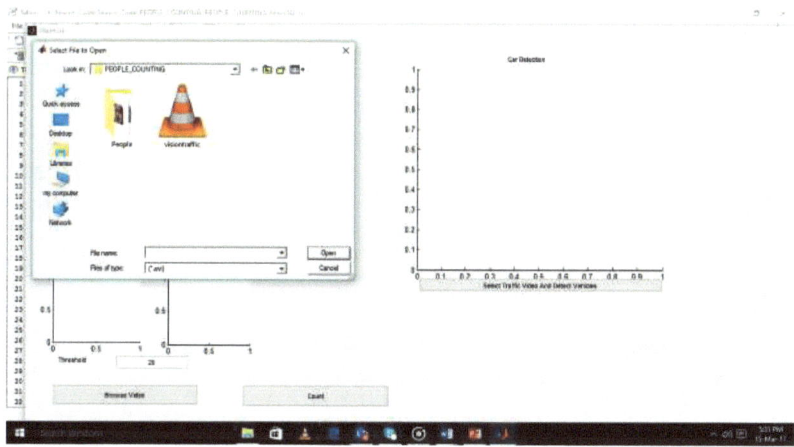

Fig. (8). Input video 2.

Foreground Image of Car

Fig. (**9**) depicts an automobile in the foreground. The foreground picture of an automobile will be displayed on the right hand image. The black colour is used for the backdrop. The white hue of the automobile distinguishes it from the surroundings.

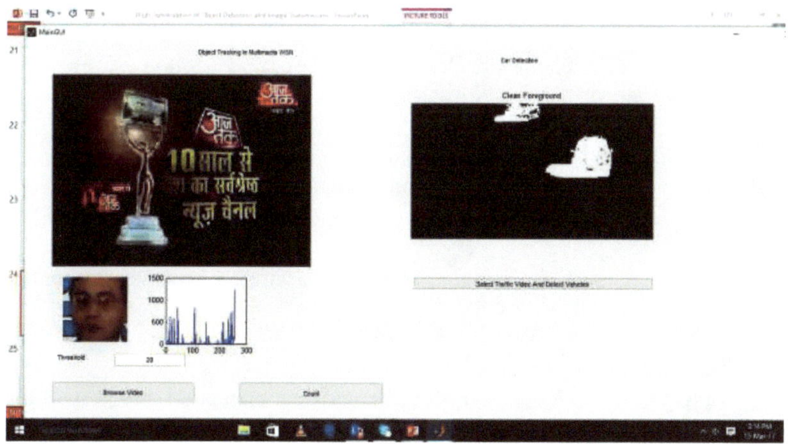

Fig. (9). Foreground image of a car.

CONCLUSION

We provided a strategy for efficient object identification and picture transmission in WMSN nodes in this research. During the transmission of visual frames, our technique conserves energy. Because of the size decrease, these photographs may be transferred using the available bandwidth. Moreover, the object presence approach has been used to quickly locate an object when it is presented on a screen despite preserving in-node energy. Furthermore, transmitting specific picture fragments improves image reconstruction at the sink node. A custom-built hardware WMSN node was used to test the method. During the reconstruction process at the sink node, the PSNR of the received picture is taken into consideration.

REFERENCES

[1]   A. Boulanouar, A. Rachedi, S. Lohier, and G. Roussel, "Energy-aware object tracking algorithm using heterogeneous wireless sensor networks", *Proc. IFIP Wireless Days* pp.1-6,2011.
[http://dx.doi.org/10.1109/WD.2011.6098160]

[2]   C. Stauffer, and W.E.L. Grimson, "Adaptive background mixture models for real-time tracking", *Proc. IEEE Comput. Soc. Conf. Comput. Vis. Pattern Recognit.* vol.02, pp.246-252,1999.
[http://dx.doi.org/10.1109/CVPR.1999.784637]

[3]   D.M. Pham, and S.M. Aziz, "Object extraction scheme and protocol for energy efficient image communication over wireless sensor networks", *Comput. Netw.,* vol. 57, no. 15, pp. 2949-2960, 2013.
[http://dx.doi.org/10.1016/j.comnet.2013.07.001]

[4]   I.F. Akyildiz, T. Melodia, and K.R. Chowdhury, "A survey on wireless multimedia sensor networks", *Comput. Netw.,* vol. 51, no. 4, pp. 921-960, 2007.
[http://dx.doi.org/10.1016/j.comnet.2006.10.002]

[5]   H. Wu, and A.A. Abouzeid, "Energy efficient distributed image compression in resource-constrained multihop wireless networks", *Comput. Commun.,* vol. 28, no. 14, pp. 1658-1668, 2005.
[http://dx.doi.org/10.1016/j.comcom.2005.02.018]

[6]   J.R. Martínez-de Dios, A. Jiménez-González, and A. Ollero, *Localization and tracking using camera-based wireless sensor networks.* InTech: Rijeka, Croatia, 2011.
[http://dx.doi.org/10.5772/19357]

[7]   L. Unzueta, M. Nieto, A. Cortés, J. Barandiaran, O. Otaegui, and P. Sánchez, "Adaptive multicue background subtraction for robust vehicle counting and classification", *IEEE Trans. Intell. Transp. Syst.,* vol. 13, no. 2, pp. 527-540, 2012.
[http://dx.doi.org/10.1109/TITS.2011.2174358]

[8]   M. Casares, M.C. Vuran, and S. Velipasalar, "Design of a wireless vision sensor for object tracking in wireless vision sensor networks", *Second ACM/IEEE International Conference on Distributed Smart Cameras.* 07-11 September,Palo Alto, CA, USA, 2008.
[http://dx.doi.org/10.1109/ICDSC.2008.4635736]

[9]   M. Chen, Q. Yang, Q. Li, G. Wang, and M-H. Yang, "Spatiotemporal background subtraction using minimum spanning tree and optical flow", In: *Computer Vision : ECCV.,* D. Fleet, T. Pajdla, B. Schiele, T. Tuytelaars, Eds., Springer: Cham, 2014, pp. 521-534.
[http://dx.doi.org/10.1007/978-3-319-10584-0_34]

[10]   M. Magno, F. Tombari, D. Brunelli, L. Di Stefano, and L. Benini, "Multimodal video analysis on self-powered resource-limited wireless smart camera", *IEEE J. Emerg. Sel. Top. Circuits Syst.,* vol. 3, no.

2, pp. 223-235, 2013.
[http://dx.doi.org/10.1109/JETCAS.2013.2256833]

[11]   M. Nasri, A. Helali, H. Sghaier, and H. Maaref, "Adaptive image transfer for wireless sensor networks (WSNs)", *Proc. 5th Int. Conf. DesignTechnol. Integr. Syst. Nanosc. Era (DTIS).* pp.1-7, 2010.

[12]   M.S. Alhilal, A. Soudani, and A. Al-Dhelaan, "Low power scheme for image based object identification in wireless multimedia sensor networks", *Proc. Int. Conf. Multimedia Comput. Syst. (ICMCS)* pp.927-932, 2014.
[http://dx.doi.org/10.1109/ICMCS.2014.6911274]

[13]   O. Ozdemir, R. Niu, and P. Varshney, "Tracking in wireless sensor networks using particle filtering: Physical layer considerations", *IEEE Trans. Signal Process.,* vol. 57, no. 5, pp. 1987-1999, 2009.

[14]   P. M. Djuric, M. Vemula, and M. F. Bugallo, "Target tracking by particle filtering in binary sensor networks", *IEEE Trans. Signal Process.,* vol. 56, no. 6, pp. 2229-2238, 2008.

[15]   P.V. Pahalawatta, T.N. Pappas, and A.K. Katsaggelos, "Optimal sensor selection for video-based target tracking in a wireless sensor network", *Proc. Int. Conf. Image Process. (ICIP),* 24-27 October,Singapore, pp.3073-3076, 2004.
[http://dx.doi.org/10.1109/ICIP.2004.1421762]

[16]   R.C. Gonzalez, *Digital image processing.* Pearson Education: India, 2009.

[17]   S. Oh, "A scalable multi-target tracking algorithm for wireless sensor networks", *Int. J. Distrib. Sens. Netw.,* vol. 8, no. 9, p. 938521, 2012.
[http://dx.doi.org/10.1155/2012/938521]

[18]   S. Yoshinaga, A. Shimada, H. Nagahara, and R. Taniguchi, "Object detection based on spatiotemporal background models", *Comput. Vis. Image Underst.,* vol. 122, pp. 84-91, 2014.
[http://dx.doi.org/10.1016/j.cviu.2013.10.015]

[19]   V. Lecuire, C. Duran-Faundez, and N. Krommenacker, "Energy-efficient transmission of wavelet-based images in wireless sensor networks", *EURASIP J. Image Video Process.,* vol. 2007, no. 1, p. 047345, 2007.
[http://dx.doi.org/10.1186/1687-5281-2007-047345]

[20]   V. Lecuire, C.D. Faundez, and N. Krommenacker, "Energy-efficient image transmission in sensor networks", *I. J. Sen. Netw.,* vol. 4, no. 1/2, pp. 37-47, 2008.
[http://dx.doi.org/10.1504/IJSNET.2008.019250]

[21]   W. Fang, D.H. Wang, and Y. Wang, "Energy-efficient distributed target tracking in wireless video sensor networks", *Int. J. Wirel. Inf. Netw.,* vol. 22, no. 2, pp. 105-115, 2015.
[http://dx.doi.org/10.1007/s10776-015-0264-1]

[22]   X.H. Fang, W. Xiong, B.J. Hu, and L.T. Wang, "A moving object detection algorithm based on color information", *J. Phys. Conf. Ser.,* vol. 48, no. 1, pp. 384-387, 2006.
[http://dx.doi.org/10.1088/1742-6596/48/1/072]

[23]   Y.M. Chi, R. Etienne-Cummings, G. Cauwenberghs, P. Carpenter, and K. Colling, "Video sensor node for low-power ad-hoc wireless networks", *Proc. 41st Annu. Conf. Inf. Sci. Syst.* 14-16 March,Baltimore, MD, USA,pp.244-247, 2007.
[http://dx.doi.org/10.1109/CISS.2007.4298307]

[24]   Y. Sugaya, and K. Kanatani, "Extracting moving objects from a moving camera video sequence", *Proc. 10th Symp. Sens. via Imag. Inf.* pp.279-283, 2004.

[25]   Y. Wang, P-M. Jodoin, F. Porikli, J. Konrad, Y. Benezeth, and P. Ishwar, "An expanded change detection benchmark dataset", *Proc. IEEE Conf. Comput. Vis. Pattern Recognit. Workshops.* pp.393-400, 2004.

[26] Z.J. Zhang, C.F. Lai, and H.C. Chao, "A green data transmission mechanism for wireless multimedia sensor networks using information fusion", *IEEE Wirel. Commun.,* vol. 21, no. 4, pp. 14-19, 2014.
[http://dx.doi.org/10.1109/MWC.2014.6882291]

CHAPTER 10

ANN Based Malicious IoT-BoT Traffic Detection in IoT Network

R. Kabilan[1,*], M. Philip Austin[1,*], J. Zahariya Gabrie[2] and Ravi R.[2]

[1] *Department of ECE, Francis Xavier Engineering College, Affiliated with Anna University, 103/G2, Bypass Road, Vannarpettai, Thirunelveli, Tirunelveli, Tamil Nadu 627003, India*

[2] *Department of Electronics and Communication Engineering, Francis Xavier Engineering College, Tirunelveli, India*

Abstract: The purpose of this study is to discover anomalies and malicious traffic in the Internet of Things (IoT) network, which is critical for IoT security, as well as to keep monitoring and stop undesired traffic flows in the IoT network. For this objective, a number of researchers have developed several machine learning (ML) approach models to limit fraudulent traffic flows in the Internet of Things network. On the other side, due to poor feature selection, some machine learning algorithms are prone to misclassifying mostly damaging traffic flows. Nonetheless, further study is needed on the vital problem of how to choose helpful attributes for accurate malicious traffic identification in the Internet of Things network. As a solution to the problem, an Artificial Neural Network (ANN) model is proposed. The Area under Curve (AUC) metric is used to employ the cross-entropy approach to effectively filter features using the confusion matrix and identify effective features for the chosen Machine Learning algorithm.

Keywords: Artificial neural network (ANN), IoT Network, Machine learning (ML), Malicious traffic.

INTRODUCTION

Large volumes of sensitive user data are vulnerable to internal and external assaults. As technology has improved, cyber-attacks have evolved in tandem with algorithm sophistication. Attacks on systems that process, store, or rely on significant data or services are the most popular targets for cyberattacks. An individual Intrusion Detection System (IDS) is necessary for the detection of destructive cyber-attacks that pose a security risk. At both the network and host

* **Corresponding authors R. Kabilan and M. Philip Austin:** Department of ECE, Francis Xavier Engineering College, Affiliated with Anna University, 103/G2, Bypass Road, Vannarpettai, Tirunelveli, Tamil Nadu 627003, India; E-mails: rkabilan13@gmail.com, philipaustin.pg20.cs@francisxavier.ac.in

levels, an intrusion detection system detects and classifies attacks, security policy violations, and intrusions [1-5].

The changing nature of threats has needed extensive modification of IDS performance, which has been accomplished by using Machine Learning (ML). Machine learning (ML) is a subfield of AI that allows computers to learn without the necessity of external programming. Machine learning algorithms aid in prediction by learning from existing data. The ultimate objective of machine learning is to develop an efficient algorithm that analyses incoming data and generates a prediction using statistical analysis.

There are two types of machine learning algorithms: 1) supervised learning and 2) unsupervised learning [6-10].

For supervised learning, a well-labelled training dataset containing both normal and attack samples is necessary. This is a type of learning in which the input and intended results are supplied to the learning model in order for it to generate future predictions. In a binary classification problem, the true or false labels must be adequate in number when each data sample includes a high number of characteristics utilised as an input to train the model. The dataset used for training may have a big influence on how well a machine-learning model works. An erroneous classification or prediction might be caused by a skewed dataset.

To attain optimal performance for accurate prediction and classification of threat types, modern IDS must use the power of AI through machine learning (ML). The datasets used to train these ML models determine the amount of training required. Biases in data or an algorithm that are overlooked or hidden might lead to skewed predictions, harming the performance of an AI application.

The Internet of Things (IoT) has become increasingly popular in recent years all across the world. By 2030, the total number of linked IoT devices is expected to reach 125 billion. The integration of these IoT devices with a variety of other technologies, services, and protocols has made IoT network management more difficult. As a result, the internet is subject to major cyber-attacks and dangers, putting users of such gadgets in peril. Distributed Denial-of-service (DDoS), DoS, ransom ware, and botnet attacks are some of the most common attacks on IoT systems.

There are various IoT-based datasets that are utilised in anomaly detection research, and the BoT-IoT dataset is one of the most well-known. The BoT-IoT dataset was created on a realistic test bed that included both simulated and real-world IoT network traffic, as well as several forms of assault. For multiclass classisification, this dataset comprises labelled features identifying the attack flow, category, and subcategories [11-18].

The BoT-IoT dataset, which was published in 2019, is the centre of this research. In terms of features and assault kinds, this dataset is extensive. The purpose of this research is to look into the factors that contribute to the dataset imbalance.

EXISTING SYSTEM

Existing technical feature selection challenges are explored and resilient features are selected by proposing several ways for Instant Message (IM) traffic identification and assault traffic detection in our existing system . Similarly, multiple feature selection strategies are given for accurate network traffic classification utilising machine learning algorithms to tackle the difficulty of feature selection. However, based on the findings of the previous study, we concluded that picking a larger range of features is inefficient for successful identification using ML approaches and that selecting a larger set of features can reduce ML classifier accuracy and increase computing complexity. However, no viable machine learning model for detecting cyber-attacks on IoT networks has yet been suggested. As a result, it is critical to investigate the problem of effective feature selection for anomaly and malicious traffic in the IoT network and present a new technique to solve it.

Support Vector Machine Algorithm

The Support Vector Machine, or SVM, is a Supervised Learning technique that may be used to solve both classification and regression issues. However, it is mostly utilised in Machine Learning for classification difficulties as shown in Fig. (1).

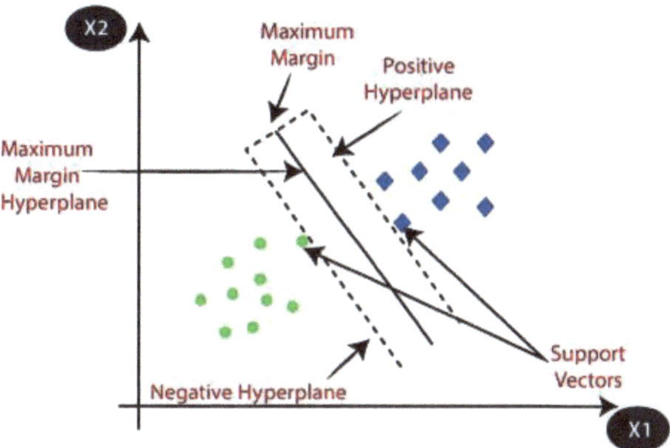

Fig. (1). Service Vector Machine.

The SVM algorithm's purpose is to find the optimum line or decision boundary for categorising n-dimensional space into classes so that additional data points can be readily placed in the correct category in the future. A hyperplane is the name for the optimal choice boundary.

Naive Bayes Classifier

A Naive Bayes Classifier is a type of classifier that is used to classify data. The Naive Bayesian model (NBN) is simple to construct and extremely useful when dealing with large datasets. Direct acyclic graphs with one parent and multiple offspring are used in this method. It implies that child nodes separated from their parents are self-contained.

Decision Trees

Decision Trees classify instances by arranging them according to their feature value. Each mode is a feature of an instance in this approach. It should be categorised, and each branch indicates a possible value for the node. It is an extensively used categorization method. In this method, classification is accomplished through the use of a decision tree.

PROPOSED SYSTEM

Many academics have provided numerous machine learning (ML) approach models to block fraudulent communication flows in the Internet of Things network. Several machine learning algorithms, on the other hand, are prone to misclassifying predominantly harmful traffic flows due to insufficient feature selection. Nonetheless, the important topic of how to pick useful features for accurate malicious traffic identification in the Internet of Things network needs to be researched further. An Artificial Neural Network model is presented as a solution to the problem. which uses a cross-entropy strategy to accurately filter features using a confusion matrix and choose effective features for the chosen Machine Learning algorithm using the AUC measure.

Data Pre-Processing in Machine Learning

Data preprocessing is the procedure for preparing raw data for use in a machine learning model. It's the first and the most important stage in building a machine learning model.

It is necessary to clean the data and format it. We use the data preprocessing task for this as shown in Fig. (**2**).

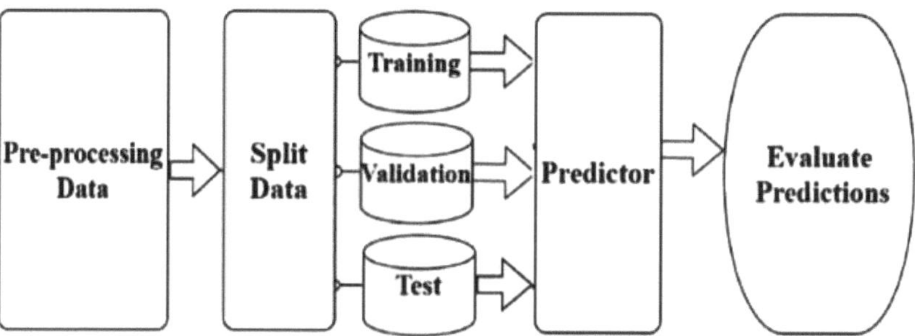

Fig. (2). Steps involved in the proposed system.

Splitting Data

Every ML problem revolves around data. ML models are like bodies without souls if they aren't fed with the right data. However, in today's era of "big data," gathering data is no longer a large issue. Every day, we consciously (or unknowingly) generate massive datasets. Having an abundance of data, on the other hand, does not address the problem. We must not only feed enormous amounts of data into ML models in order for them to produce respectable results, but we must also ensure that the data is of high quality.

Training Data

The data scientist feeds the algorithm input data, which corresponds to an expected output. The model evaluates the data repeatedly to learn more about the data's behaviour and then adjusts itself to serve its intended purpose.

Validation Data

Although understanding raw data is an art in and of itself, requiring high feature engineering skills and domain expertise (in some cases), quality data is useless unless it is used correctly. The most challenging problem that ML/DL practitioners encounter is partitioning data for training and testing. Although it appears to be a straightforward problem at first glance, only a thorough examination reveals its full complexity. The model's output can be unexpected if the training and testing sets are inadequate. It could lead to data overfitting or underfitting, leading to skewed results from our model.

Test Data

Testing data once the model has been developed confirms that it can make accurate predictions. The testing data should be unlabeled if the training and validation data include labels to track the model's performance metrics. Test data is a last real-world verification of an unknown dataset to ensure that the machine learning algorithm was properly trained.

Predicator

The amount of data used is exponentially expanding, and today, a vast volume of big data is accumulated across businesses, which could be related to business partners, consumers, application partners, internal and external executives, visitors, and so on. To detect and analyse trends, data are churned and categorised. Data Analytics, on the other hand, is a process that uses a variety of tools and techniques for qualitative and quantitative research to use this accumulated data and provide certain outcomes that are used to improve performance, yield, risk reduction, and business productivity.

Artificial Neural Networks

It was designed to mimic the behaviour of biological systems made up of "neurons." It has machine learning and pattern recognition capabilities. These were shown as interconnected "neurons" that could compute values based on inputs is shown in Fig. (**3**).

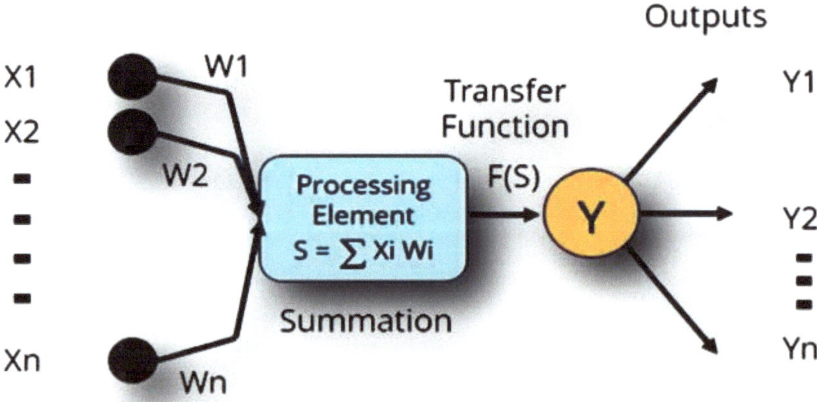

Fig. (3). Basic Block diagram of ANN.

It works in the same way as the human brain does. ANN is made up of a large number of interconnected processing units that collaborate to process data. They also produce useful outcomes as a result of it. It can be used for continuous target attribute regression.

$$f(x) = b_0 + \sum_{i=0}^{n} (b_i * W_i) \tag{1}$$

b_i - input W_i -Weights, b_0– bias

In the field of data mining, neural networks have a lot of applications. For example, pattern recognition in economics, forensics, and other fields. After thorough training, it can also be utilised for data classification in vast amounts of data.

Artificial Neural Network Layers

In most cases, an artificial neural network is organised in layers. Layers are made up of a series of interconnected 'nodes' that each have an 'activation function.' is shown in Fig. **(4)**.

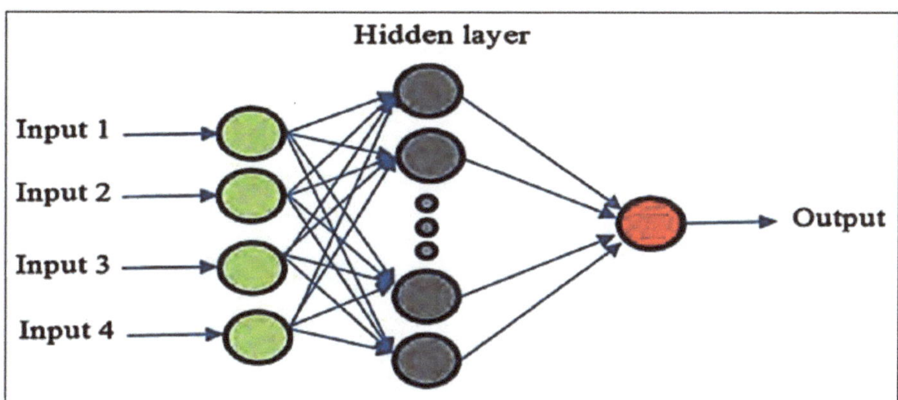

Fig. (4). ANN Layer.

Input Layer

The input layer's job is to take in the values of each observation's explanatory characteristics as input. An input layer's number of input nodes is usually equal to the number of explanatory variables. The patterns are presented to the network through the 'input layer,' which communicates with one or more 'hidden layers.'

The input layer's nodes are passive, which means they don't affect the data. They take a single value from their input and duplicate it across all of their outputs. It repeats each value from the input layer and sends it to all hidden nodes.

Hidden Layer

The hidden layers inside the network apply alterations to the input values. Incoming arcs from other hidden nodes or input nodes connected to each node are included in this. It connects to output nodes or other hidden nodes *via* outgoing arcs. The real processing is done in the hidden layer *via* a system of weighted 'connections.' There could be one or more layers buried beneath the surface. Weights, a set of predetermined numbers contained in the programme, are multiplied by the values entering a hidden node. After that, the weighted inputs are summed together to form a single number.

Output Layer

It returns a value that corresponds to the prediction of the response variable. The active nodes in the output layer combine and manipulate data to create output values. The ability of the neural network to provide useful data manipulation depends on the weights being selected correctly. This differs from traditional data processing.

Structure of a Neural Network

The 'architecture' or 'topology' of a neural network is another term for its structure. Interconchangend Weight Adjustment Mechanism is also included. The structure that is chosen determines the outcomes that will be obtained. It is the most important aspect of a neural network's implementation.

Units are distributed in two layers: an input layer and an output layer. This is the most basic arrangement. The input layer's units have a single input and a single output that is the same as the input. With a combination function and a transfer function, the output unit has all units of the input layer connected to its input. One or more output units may be present. A linear or logistic regression is the resulting model in this scenario. Whether the transfer function is linear or logistic determines this.

The network's weights are regression coefficients. The predictive potential of a neural network is increased by adding one or more hidden layers between the input and output layers, as well as units in this layer. This prevents overfitting by ensuring that the neural network does not keep all information from the learning set but can generalise it.

It is possible that you'll be overfitted. When weights cause the system to learn details of the learning set instead of discovering structures, this happens. This occurs when the size of the learning set is too small in comparison to the model's complexity. Whether or not a hidden layer is present, the network's output layer can sometimes have many units when there are numerous classes to forecast.

Activation Function

Sigmoid () Function

The Sigmoid activation function, often known as the logistic function, is a neural network activation function is shown in Fig. (**5**).

$$g(z) \quad \frac{1}{1 + e^{-z}} \tag{2}$$

Assume a threshold and...

Predict y = 1 if $h_\theta(x) \geq 0.5$

Predict y = 0 if $h_\theta(x) < 0.5$

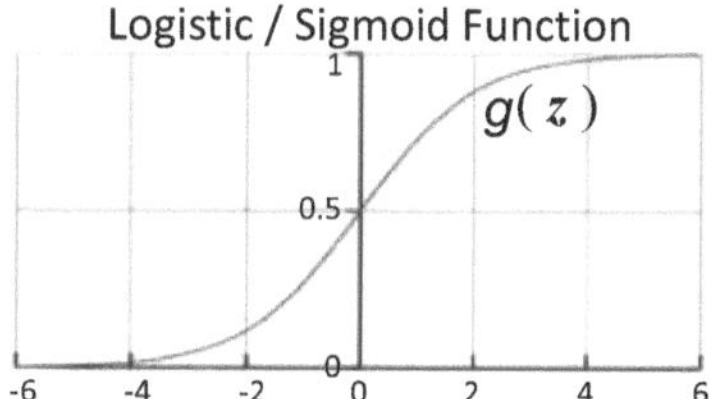

Fig. (5). Sigmoid Model.

Relu () Function

The rectified linear activation function solves the vanishing gradient problem, allowing the models to train quicker and perform better is shown in Fig. (**6**).

$$F(x) = \max(0,x) \tag{3}$$

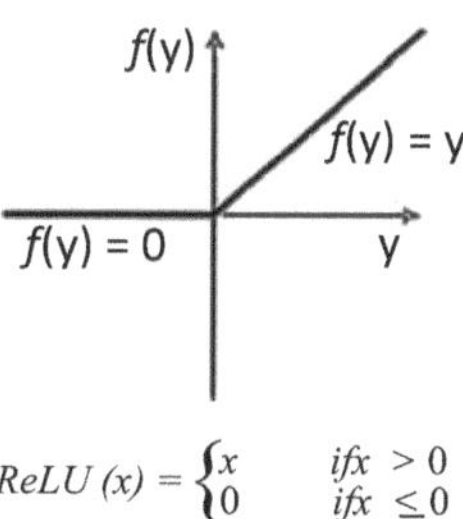

Fig. (6). ReLU Model.

Cost Function

It computes the error by calculating the difference between the actual and projected numbers.

$$C = \frac{\Sigma \, (y - a)^2}{n} \tag{4}$$

It has one disadvantage: it slows down neural network learning; hence, in this case, we employ "Cross Entropy."

Cross Entropy

It improves neural network learning by lowering the cost function.

$$C = \frac{-1}{n} * \sum_{i=0}^{n} (y * \ln(a) + (1 - y) * \ln(1 - a)) \tag{5}$$

Gradient Descent

It is a cost-cutting optimization algorithm that finds the function's minimum value is shown in Fig. **(7)**.

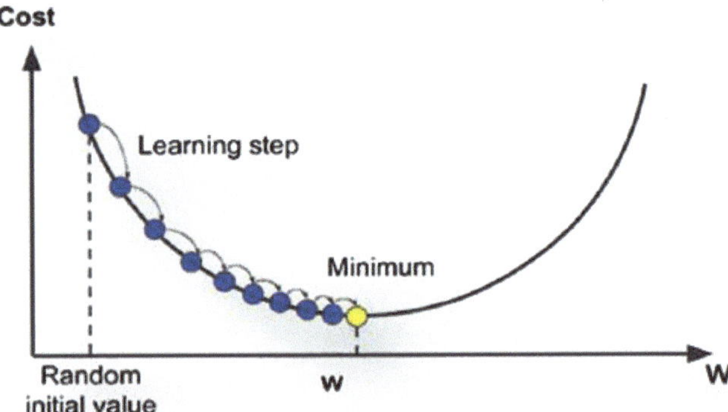

Fig. (7). Gradient descent.

PERFORMANCE METRICS

Confusion Matrix

The Confusion Matrix is a machine learning classification performance evaluation tool. True values from the testing data are used to measure performance. Confusion Matrix aids in the detection of errors (FP and FN) as well as the calculation of other performance indicators in machine learning. For binary classifiers, the confusion matrix in Table **1** comprises four elements:

Table 1. Confusion matrix.

Confusion Matrix		Actual Values	
		Positive	**Negative**
Predicted Values	Positive	TP	FP
	Negative	FN	TN

The confusion matrix's four values (TP, TN, FN, and FP) are used to evaluate the performance metrics.

DATA

Data are collected from Files – CloudStor (aarnet.edu.au). The name of the database is IoT-Bot dataset.

The raw network packets (Pcap files) of the BoT-IoT dataset were created by the application of the tshark tool in the Cyber Range Lab of the Australian Centre for Cyber Security (ACCS), and incorporates a combination of normal and abnormal traffic. Simulated network traffic was generated through Ostinato tool and Node-red (for non-IoT and IoT respectively). The dataset's source files are provided in different formats, such as the original pcap files, the generated argus files, and finally in CSV format. The files were separated based on attack category and subcategory to better assist in the labelling process.

Free use of the Bot-IoT dataset for academic research purposes is hereby granted in perpetuity. Its use for commercial purposes is strictly prohibited. Nickolaos Koroniotis, Nour Moustafa, and Elena Sitnikova have asserted their rights under the Copyright.

RESULT

Fig. (**8**) depicts the model training sequence, which changes at each epoch. It also displays the loss and accuracy for each epoch.

The maximum epoch value is set to 70. For load-free model training, the data is divided by 20. This model's accuracy is relatively close to 93 percent, and its loss is closer to 0.02 percent.

With the help of an Adam optimizer, the model is trained using an ANN model, which is a 5-layer network. The training sequence accuracy for each epoch of our model is shown in the image above (Fig. **9**).

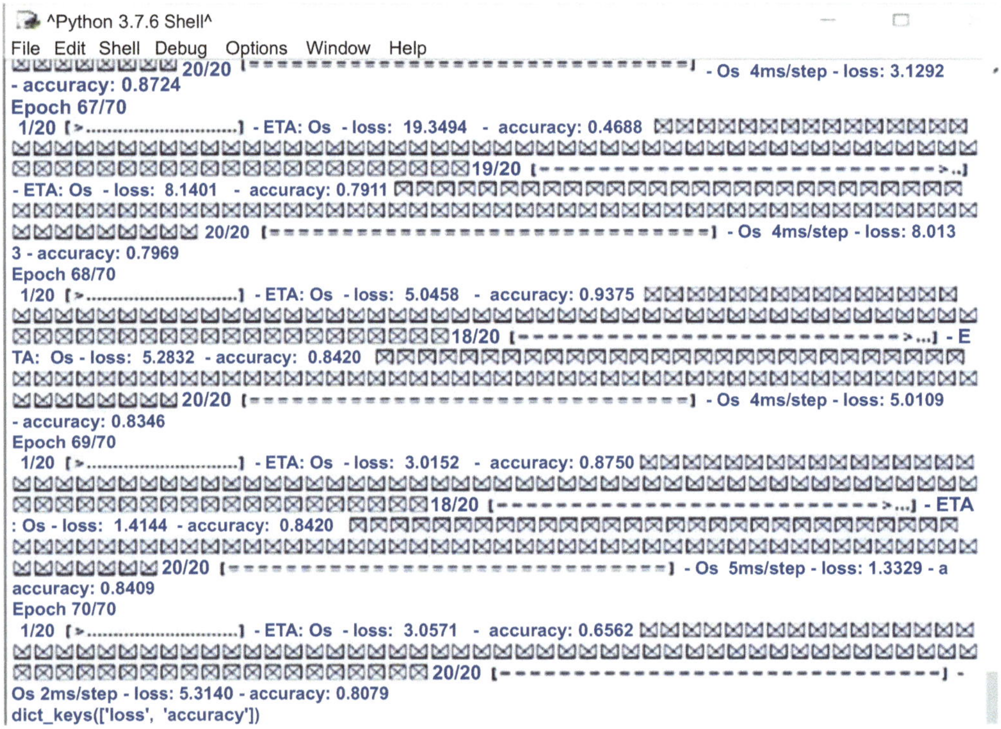

```
^Python 3.7.6 Shell^                                              —    □
File  Edit  Shell  Debug  Options  Window  Help
20/20 [==============================] - 0s  4ms/step - loss: 3.1292
- accuracy: 0.8724
Epoch 67/70
 1/20 [>.............................] - ETA: 0s - loss: 19.3494  - accuracy: 0.4688
19/20 [----------------------------->..]
- ETA: 0s - loss: 8.1401  - accuracy: 0.7911
20/20 [==============================] - 0s  4ms/step - loss: 8.013
3 - accuracy: 0.7969
Epoch 68/70
 1/20 [>.............................] - ETA: 0s - loss: 5.0458  - accuracy: 0.9375
18/20 [----------------------------->...] - E
TA: 0s - loss: 5.2832 - accuracy: 0.8420
20/20 [==============================] - 0s  4ms/step - loss: 5.0109
- accuracy: 0.8346
Epoch 69/70
 1/20 [>.............................] - ETA: 0s - loss: 3.0152  - accuracy: 0.8750
18/20 [----------------------------->...] - ETA
: 0s - loss: 1.4144 - accuracy: 0.8420
20/20 [==============================] - 0s  5ms/step - loss: 1.3329 - a
accuracy: 0.8409
Epoch 70/70
 1/20 [>.............................] - ETA: 0s - loss: 3.0571  - accuracy: 0.6562
20/20 [------------------------------] -
0s 2ms/step - loss: 5.3140 - accuracy: 0.8079
dict_keys(['loss', 'accuracy'])
```

Fig. (8). Training Data.

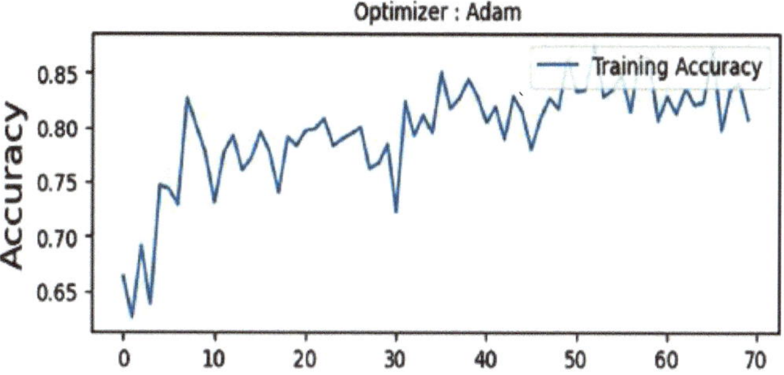

Fig. (9). Model Accuracy.

The RMSE loss of the training model is shown in Fig. (**10**) for each epoch. The training loss for each training sequence in terms of epochs is shown in the graph. As soon as the model learns the data pattern, the loss diminishes (Fig. **11**).

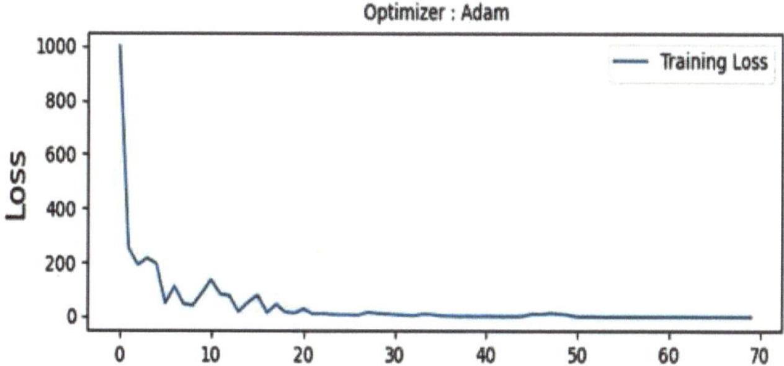

Fig. (10). Model Loss.

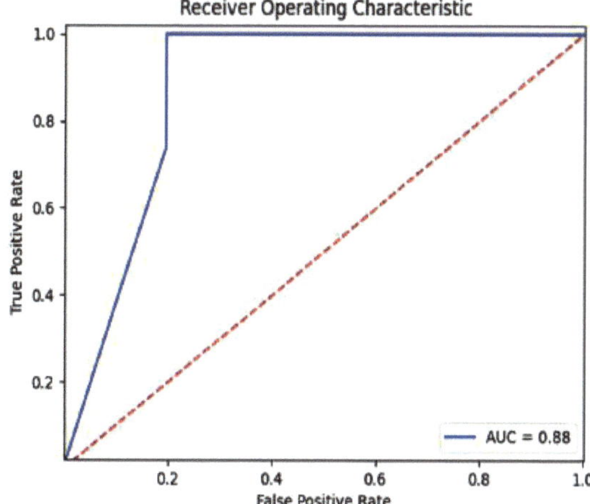

Fig. (11). AUC-ROC Curve.

The area under the ROC curve is referred to as the AUC. The higher the AUC, the more accurate the model is in classifying cases. The ROC curve should, in theory, extend to the upper left corner. In that case, the AUC score would be 1 (Fig. **12**).

```
Accuracy: 0.899371
Precision: 0.827957
Recall: 1.000000
F1 score: 0.905882

ROC AUC: 0.875831
Confussion Matrix
[[66 16]
 [ 0 77]]
```

Fig. (12). Performance metrics.

Above results were found with the help of Table **1** as shown in Fig. (**13**).

```
[794 rows x 19 columns]
<class 'pandas.core.frame.DataFrame'>
Int64Index: 794 entries, 0 to 793
Data columns (total 19 columns):
 #   Column             Non-Null Count  Dtype
---  ------             --------------  -----
 0   pkSeqID            794 non-null    int64
 1   proto              794 non-null    object
 2   saddr              794 non-null    object
 3   sport              794 non-null    object
 4   daddr              794 non-null    object
 5   dport              794 non-null    object
 6   seq                794 non-null    int64
 7   stddev             794 non-null    float64
 8   N_IN_Conn_P_SrcIP  794 non-null    int64
 9   min                794 non-null    float64
 10  state_number       794 non-null    int64
 11  mean               794 non-null    float64
 12  N_IN_Conn_P_DstIP  794 non-null    int64
 13  drate              794 non-null    float64
 14  srate              794 non-null    float64
 15  max                794 non-null    float64
 16  attack             794 non-null    int64
 17  category           794 non-null    object
 18  subcategory        794 non-null    object
dtypes: float64(6), int64(6), object(7)
memory usage: 124.1+ KB
```

Fig. (13). Features selected from IoT-BoT data set.

A request is shown in Figs. (**14** and **15**) depicts an application for putting the ANN-trained model to the test a data request. We can determine if the request is legitimate or malicious by receiving it and analysing the data. The output has been identified as a malicious request in this instance.

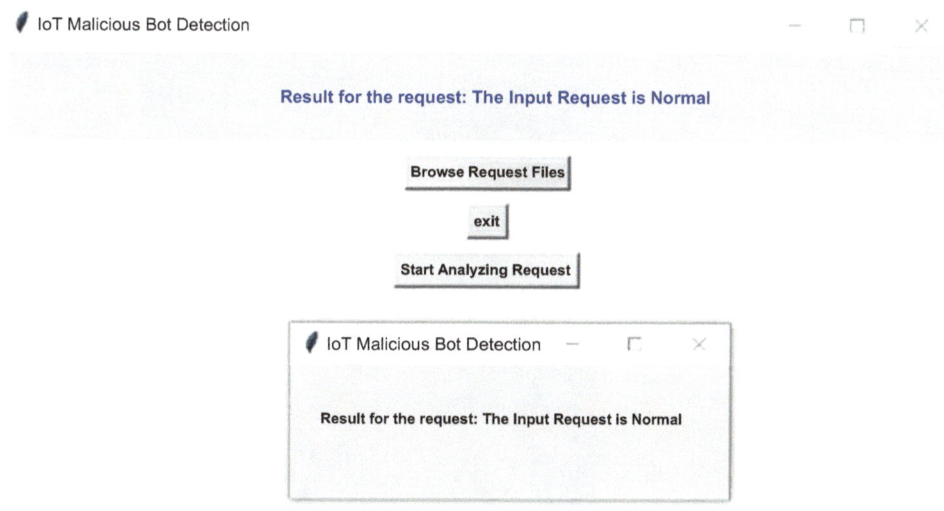

Fig. (14). A request is attached to the application.

Fig. (15). Request 1 is tested with the trained model.

Fig. (**16**) depicts an application for putting the ANN-trained model to the test a data request. We can determine if the request is legitimate or malicious by receiving it and analysing the data. The output has been identified as a legitimate request in this instance. The app was developed with Tinker Python package is shown in Fig. (**17**).

Fig. (16). Request 1 is tested with the trained model.

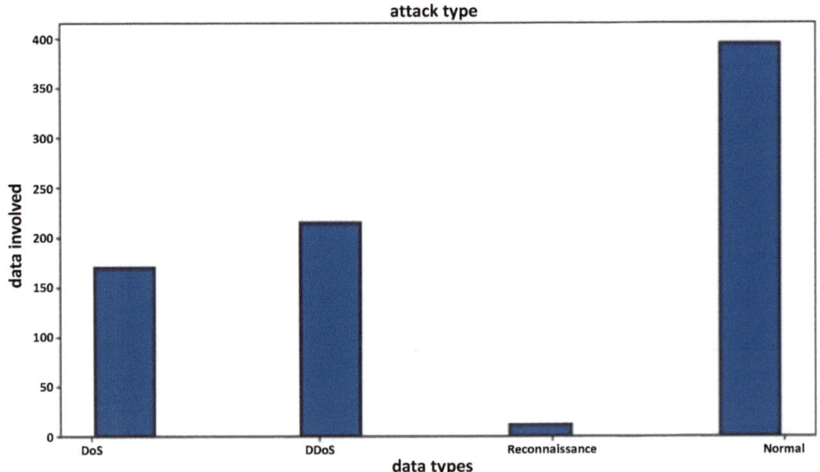

Fig. (17). Type of data involved.

The graph depicts the number of data points used in DoS and DDoS attacks, as well as the number of normal data points used to train our model.

Fig. (**18**) depicts the different types of procedures used in relation to the amount of data in our dataset. TCP, UDP, and ICMP are the protocols involved.

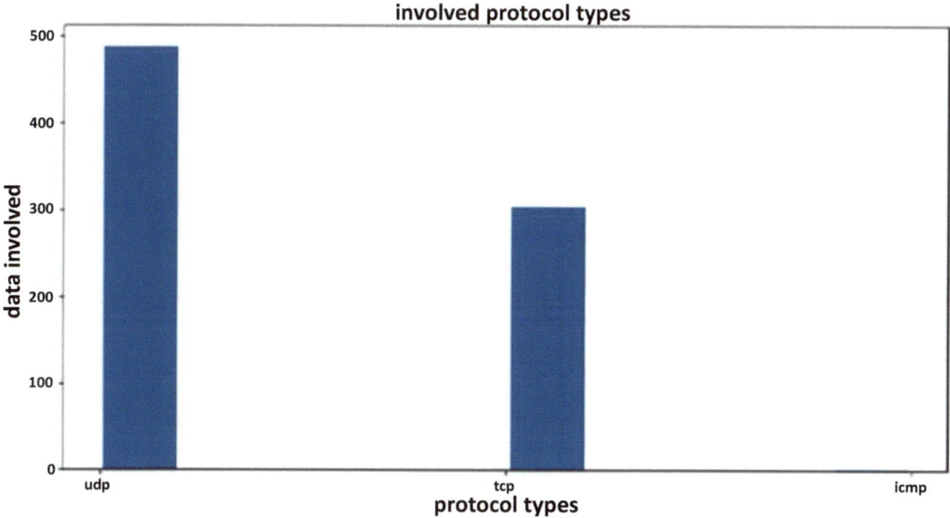

Fig. (18). Protocols involved.

CONCLUSION

The topic of discussion in this study was ANN-Based Malicious IoT-BoT Traffic Detection IN IoT Networks. This study uses the IoT-BoT dataset to train the machine model and estimate the network trustworthiness of IoT devices. The model is given a set of criteria to predict whether an IoT bot request is trustworthy or malicious. In this study, one of the deep learning methodologies, Artificial Neural Network, is employed to predict the result. Future research may employ hybrid Deep Learning techniques to boost forecast accuracy. This research was made feasible using Python 3.7, numpy, pandas, Google Tensor flow tools, and Tinker tools.

REFERENCES

[1]   M.A. Alsheikh, S. Lin, D. Niyato, and H.P. Tan, "Machine learning in wireless sensor networks: Algorithms, strategies, and applications", *IEEE Commun. Surv. Tutor.*, vol. 16, no. 4, pp. 1996-2018, 2014.
[http://dx.doi.org/10.1109/COMST.2014.2320099]

[2]   D.E. Denning, "An intrusion-detection model", *IEEE Trans. Softw. Eng.*, vol. SE-13, no. 2, pp. 222-232, 1987.
[http://dx.doi.org/10.1109/TSE.1987.232894]

[3]   Xiaojiang Du, and Hsiao-Hwa Chen, "Security in wireless sensor networks", *IEEE Wirel. Commun.,* vol. 15, no. 4, pp. 60-66, 2008.
[http://dx.doi.org/10.1109/MWC.2008.4599222]

[4]   X. Du, M. Guizani, Y. Xiao, and H. Chen, "Defending dos attacks on broadcast authentication in wireless sensor networks in 2008", *IEEE International Conference on Communications.* 19-23 May, Beijing, China, pp.1653-1657, 2008.

[5]   S. Egea, A. Rego Manez, B. Carro, A. Sánchez-Esguevillas, and J. Lloret, "Intelligent IoT traffic classification using novel search strategy for fast based correlation feature selection in industrial environments", *IEEE Internet Things J.,* vol. 5, no. 3, pp. 1616-1624, 2018.
[http://dx.doi.org/10.1109/JIoT.2017.2787959]

[6]   J. Qiu, Z. Tian, C. Du, Q. Zuo, S. Su, and B. Fang, "A survey on access control in the age of internet of things", *IEEE Internet Things J.,* vol. 7, no. 6, pp. 4682-4696, 2020.
[http://dx.doi.org/10.1109/JIoT.2020.2969326]

[7]   M. Shafiq, Z. Tian, A. Bashir, R. Jolfaei, and X. Yu, "Data mining and machine learning methods for sustainable smart cities traffic classification: A surve", *Sustain Cities Soc.,* vol. 60, p. 102177, 2020.
[http://dx.doi.org/10.1016/j.scs.2020.102177]

[8]   M. Shafiq, Z. Tian, Y. Sun, X. Du, and M. Guizani, "Selection of effective machine learning algorithm and Bot-IoT attacks traffic identification for internet of things in smart city", *Future Gener. Comput. Syst.,* vol. 107, pp. 433-442, 2020.
[http://dx.doi.org/10.1016/j.future.2020.02.017]

[9]   S. Su, Y. Sun, X. Gao, J. Qiu, and Z. Tian, "A correlation-change based feature selection method for iot equipment anomaly detection", *Appl. Sci.,* vol. 9, no. 3, p. 437, 2019.
[http://dx.doi.org/10.3390/app9030437]

[10]   Z. Tian, X. Gao, S. Su, and J. Qiu, "A novel reputation framework for identifying denial of traffic service in internet of connected vehicles", *IEEE Internet Things J.,* vol. 7, no. 5, pp. 3901-3909, 2020.
[http://dx.doi.org/10.1109/JIoT.2019.2951620]

[11]   Z. Tian, X. Gao, S. Su, J. Qiu, X. Du, and M. Guizani, "Evaluating reputation management schemes of internet of vehicles based on evolutionary game theory", *IEEE Trans. Vehicular Technol.,* vol. 68, no. 6, pp. 5971-5980, 2019.
[http://dx.doi.org/10.1109/TVT.2019.2910217]

[12]   Z. Tian, C. Luo, J. Qiu, X. Du, and M. Guizani, "A distributed deep learning system for web attack detection on edge devices", *IEEE Trans. Industr. Inform.,* vol. 16, no. 3, pp. 1963-1971, 2020.
[http://dx.doi.org/10.1109/TII.2019.2938778]

[13]   Z. Tian, W. Shi, Y. Wang, C. Zhu, X. Du, S. Su, Y. Sun, and N. Guizani, "Real time lateral movement detection based on evidence reasoning network for edge computing environment", *IEEE Trans. Industr. Inform.,* vol. 15, no. 7, pp. 4285-4294, 2019.
[http://dx.doi.org/10.1109/TII.2019.2907754]

[14]   D. Ventura, "Ariima: A real iot implementation of a machine-learning architecture for reducing energy consumption", *International conference on ubiquitous computing and ambient intelligence.* December 2-5,Belfast, UK,pp. 444-451, 2014.
[http://dx.doi.org/10.1007/978-3-319-13102-3_72]

[15]   L. Wu, X. Du, W. Wang, and B. Lin, "An out-of-band authentication scheme for internet of things using blockchain technology", *International Conference on Computing, Networking and Communications (ICNC),* >05-08 March,Maui, HI, USA,pp. 769-773, 2018.
[http://dx.doi.org/10.1109/ICCNC.2018.8390280]

[16]   Y. Xiao, X. Du, J. Zhang, F. Hu, and S. Guizani, "Internet protocol television (IPTV): The killer application for the next-generation internet", *IEEE Commun. Mag.,* vol. 45, no. 11, pp. 126-134, 2007.
[http://dx.doi.org/10.1109/MCOM.2007.4378332]

[17]   R. Xue, L. Wang, and J. Chen, "Using the IoT to construct ubiquitous learning environment", *Second International Conference on Mechanic Automation and Control Engineering.* 15-17 July,Hohhot, pp.7878-7880, 2011.
[http://dx.doi.org/10.1109/MACE.2011.5988881]

[18]   Y. Xiao, V.K. Rayi, B. Sun, X. Du, F. Hu, and M. Galloway, "A survey of key management schemes in wireless sensor networks", *Comput. Commun.,* vol. 30, no. 11-12, pp. 2314-2341, 2007.
[http://dx.doi.org/10.1016/j.comcom.2007.04.009]

CHAPTER 11

High-Performance Mixed Signal VLSI Design For Multimode Demodulator

R. Kabilan[1,*], J. Zahariya Gabrie[2], Ravi R.[2] and **M. Philip Austin[1]**

[1] *Department of ECE, Francis Xavier Engineering College, Affiliated with Anna University, 103/G2, Bypass Road, Vannarpettai, Tirunelveli, Tamil Nadu 627003, India*

[2] *Department of Electronics and Communication Engineering, Francis Xavier Engineering College, Tirunelveli, India*

Abstract: A mixed signal quadrature demodulator was suggested in this study. In 90 nm CMOS technology, to get the desired frequency range, a quadrature VCO is employed. The fast speed is achieved with a three-bit ADC. Unused ADC construction components have been removed to conserve energy and space. Outputs obtained are used to meet the power needed in the mixed signal demodulator designed for multi-gigabit applications. QVCO, baseband AGC, frequency synthesizers, and IQ mixers, are all part of the demodulator. This displays the highest level of integration while using the least amount of electricity. To sample the symbols at optimal SNR, the baseband modem included a mixed signal timing recovery loop based on the Gardner timing error detector.

Keywords: AGC, CMOS, Error detector, PLL, VCO.

INTRODUCTION

VLSI technology may be used to create mixed analogue digital chips with over one million transistors and other devices. We should emphasize that interfacing to the analogue world entails not only A/D or D/A converters, but also pre-and post-conversion signal conditioning (amplification, filtering for anti-aliasing or smoothing, sampling, holding, multiplexing, demultiplexing, and so on), as well as direct signal processing without conversion into digital signals, particularly in combinations with medium signal to noise ratio with high speed and or low power. Sensors, receiving antennas, transmission lines, and other circuits all contribute to the signal that must be processed by the analogue component of the MAD chips. Activators, transmitting antennas, transmission lines, and other circuits must all be driven by them. Due to substantial work on sensors and actu-

* **Corresponding author R. Kabilan:** Department of ECE, Francis Xavier Engineering College, Affiliated with Anna University, 103/G2, Bypass Road, Vannarpettai, Tirunelveli, Tamil Nadu 627003, India; E-mail: rkabilan13@gmail.com

S. Kannadhasan, R. Nagarajan, Alagar Karthick, K.K. Saravanan & Kaushik Pal (Eds.)

ators, the range of both sources and destinations for analogue signals is growing. The growing use of sensors in current systems implies that MAD circuits will play a larger role in these systems. Because of the advantages that MOS technology provides for digital circuits, the majority of MAD chips have been implemented in this technology. All of them are now on the MAD chip in modern disc drives. Another area where MAD chips are utilised is in digital communication networks for signal processing of digital pulses.

For mixed-signal applications, a digital-to-analog converter (DAC) functions as a technological test vehicle. Current steering D/A converters, which are based on an array of matched current sources that are switched to the output according to a digital input word, are commonly employed for signal processing applications. The use of both unary and binary coded bits results in a design that is both simple and resilient, as well as having a high degree of precision. The total of the binary scaled currents is obtained from the unary scaled current sources to avoid a mismatch between the binary and unary parts. These values describe performance constraints in dynamic DACs, such as harmonic distortion, synchronization and timing problems in switch control signals, or capacitive feed through the clock and data signals to the output.

Modern communication systems make extensive use of phase-locked-loop circuits. For clock creation or regeneration as a frequency synthesizer, PLLs are made up of digital (PFD & FD), analogue (CP & LF), and RF (VCO) components. The phase of the VCO is changed *via* negative feedback to match the phase of the reference clock f ref, resulting in a constant skew between the two PFD inputs. When the PLL is locked in a steady state, the desired synchronization f out = N f ref is accomplished. PLL performance is defined by power consumption, circuit size, and jitter, in addition to the goal frequency range (f ref &f out) and locking time. Jitter is the integrated phase noise of the carrier f out and describes the timing uncertainty of the clock edges. As a result, a charge-pump PLL is used as a secondary test vehicle for complicated mixed-signal circuits with a focus on noise and matching [1-6].

Existing Systems

Gilbert Cell Mixer

The Gilbert cell mixer, sometimes known as the Gilbert cell multiplier, is a kind of double-balanced mixer that uses a symmetrical architecture to eliminate undesirable RF and LO output signals from the IF.

Ring Vco

It can create multiple-phase clock signals that are utilized to sample incoming data, ring VCOs are frequently employed in clock and data recovery phase-locked loops. Ring oscillators are also the easiest way to evaluate a technology's performance. Many publications on CML ring oscillators employing SiGe or GaAs technologies have already been published, but only a few papers have documented success in deep sub-micron silicon CMOS technology.

Adc

A device that transforms a continuous number to a discrete digital number is known as an analog-to-digital converter (abbreviated ADC, A/D, or A to D). An ADC is a type of electrical device that transforms an analogue voltage (or current) into a digital number proportionate to its magnitude. Different coding techniques may be used in the digital output is shown in Fig. (**1**).

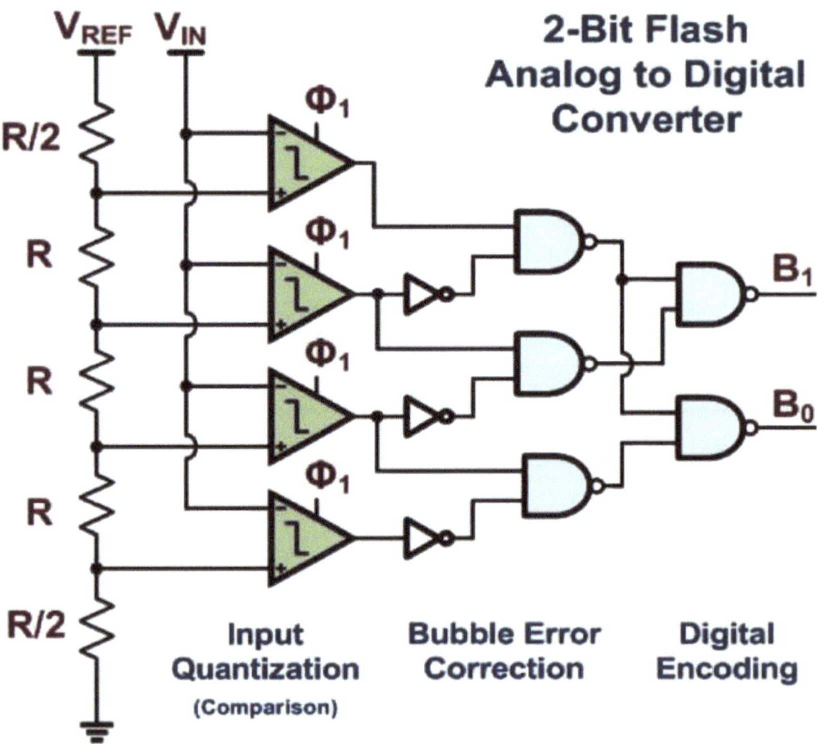

Fig. (1). Two-bit flash ADC.

Successive Approximation Register

An approximation that is made in ADCs are analog-to-digital converters that use a binary search to transform a continuous analogue waveform into a discrete digital representation before eventually settling on a digital output for each conversion (Fig. **2**).

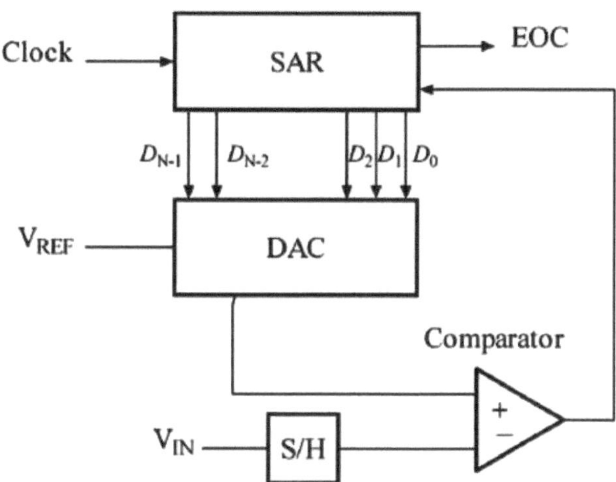

Fig. (2). Successive approximation ADC block diagram.

Charge-Redistribution Successive Approximation ADC

A charge scaling DAC is used in one of the most prevalent versions of the successive approximation ADC, *i.e.* the charge-redistribution successive approximation ADC is shown in Fig. (**3**).

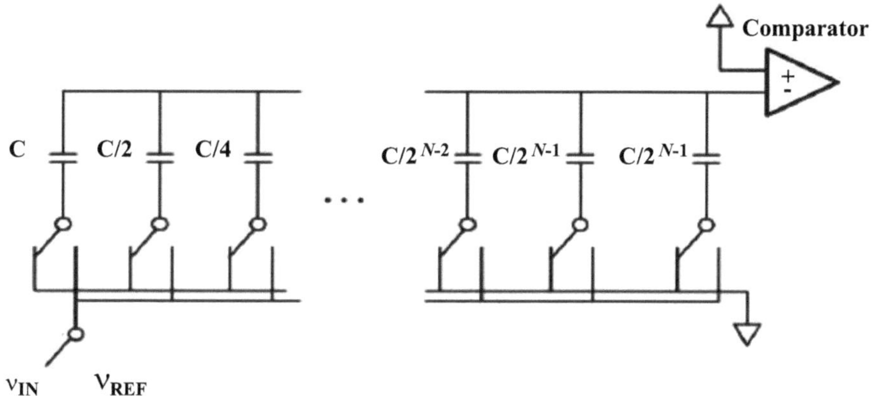

Fig. (3). Charge scaling DAC.

PROPOSED SYSTEM

Modulation Scheme

The inherent and extrinsic noise sources determine the needed performance of a wireless receiver.

In relation to this information, the receiver's minimum sensitivity may be represented as,

$$P_{min}=10 \log(kT.BW)+NFR+LLM+SNR_{min} \qquad (2)$$

where K denotes the Boltzmann constant, T is the absolute temperature in Kelvin, BW is the RF signal's bandwidth, NFR is the receiver's noise figure, LLM is the link margin, and SNR min is the lowest signal-to-noise ratio (SNR) necessary for the modulation chosen. The baseband SNR necessary for a particular modulation scheme may be determined to get a certain degree of reliability in terms of (BER).

$$SNR_{ADC}=P_{in}-10 \log(kT.BW)-NF \qquad (3)$$

Receiver Architecture

The quadrature demodulator down changes the modulated IF carrier to baseband using double-balanced passive mixers (Fig. **4**).

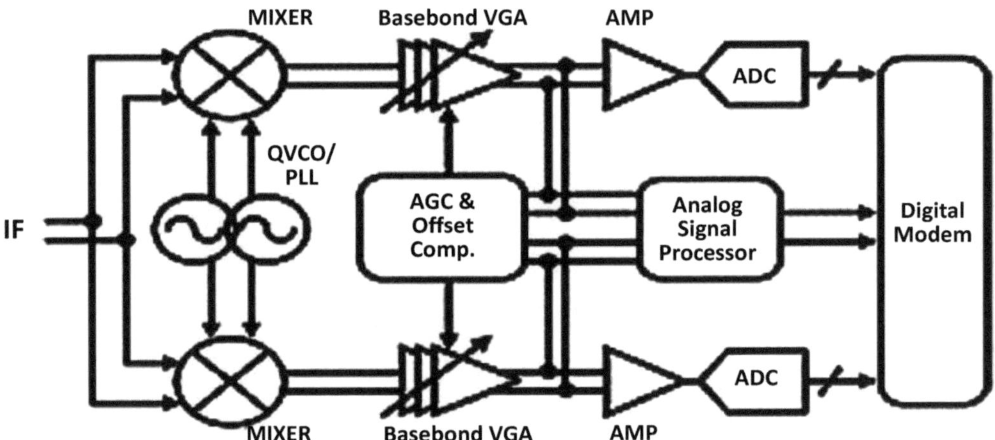

Fig. (4). Demodulator block diagram.

CIRCUIT IMPLEMENTATION

Frequency Synthesizer

The QVCO is locked to a frequency of 13GHz using the PLL (Fig. **5**). This PLL has a fourth-order Integer-N type-II design. A customizable division ratio allows for different data speeds (432MHz, 864MHz, 1.485GHz, and 1.728GHz) (Fig. **6**).

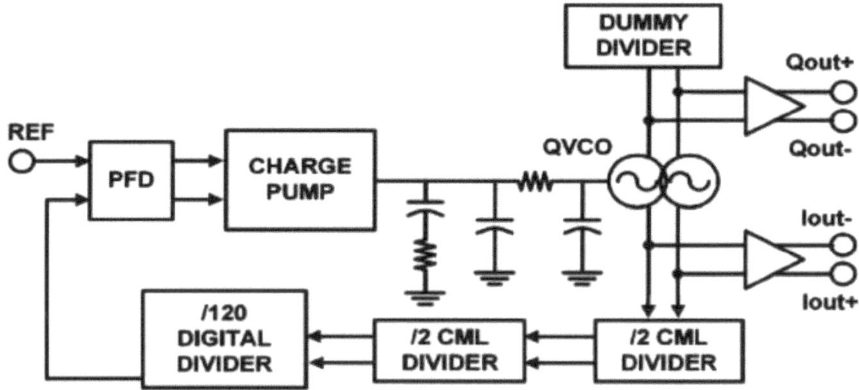

Fig. (5). Block diagram of frequency synthesizer.

IQ Mixers

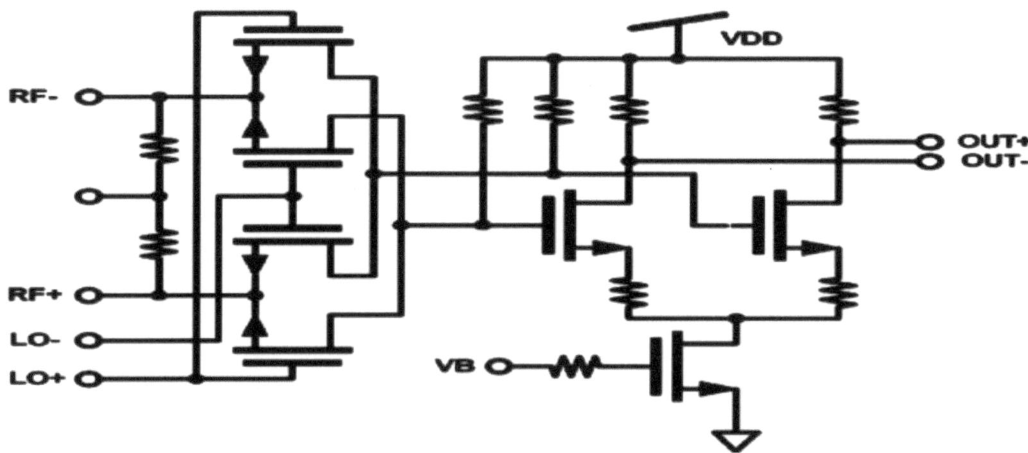

Fig. (6). Schematic diagram representation of double-balanced passive mixer.

High-Speed Analog-to-Digital Converter

The input is compared to a voltage reference, V ref, in this CMOS inverter circuit utilizing the differential pair at output is shown in Fig. (**7**).

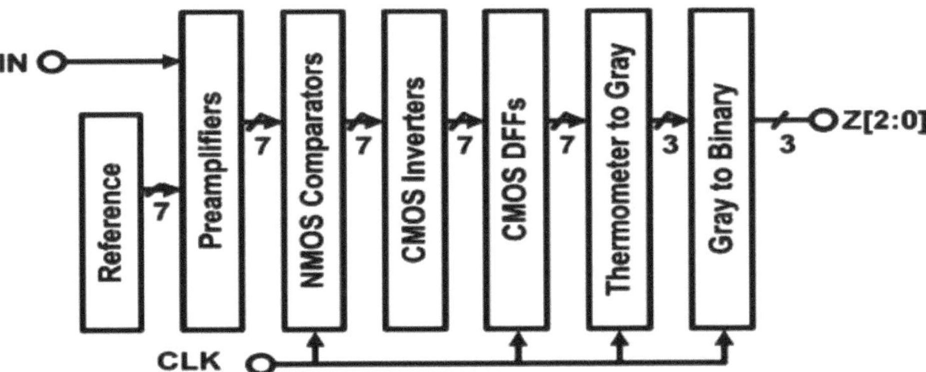

Fig. (7). Block diagram of high speed 3 bit ADC.

Mixed Signal Back End

The charge pump which obtains 300 A, and its loop bandwidth has been set to the 2 MHz range to prevent the use of a big external capacitor is shown in Fig. (**8**).

The charge pump is shown in Fig. (**9**).

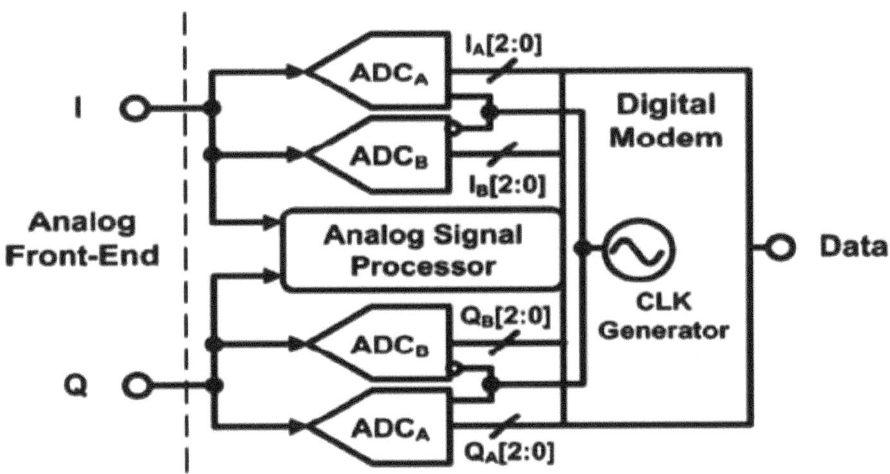

Fig. (8). Block diagram of mixed signal back end.

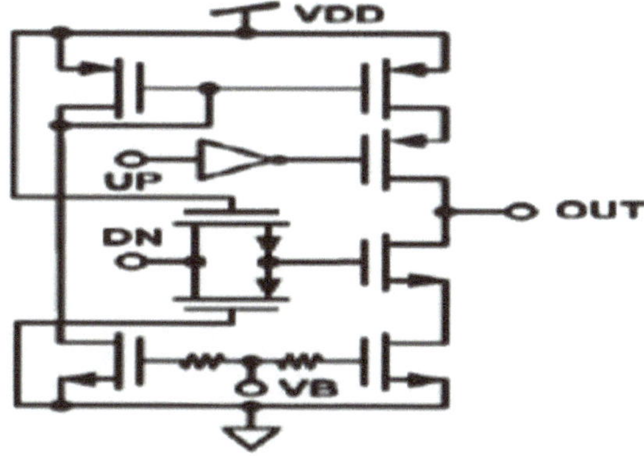

Fig. (9). Charge pump diagram.

RESULTS AND DISCUSSIONS

1) Input waveform (sine wave)

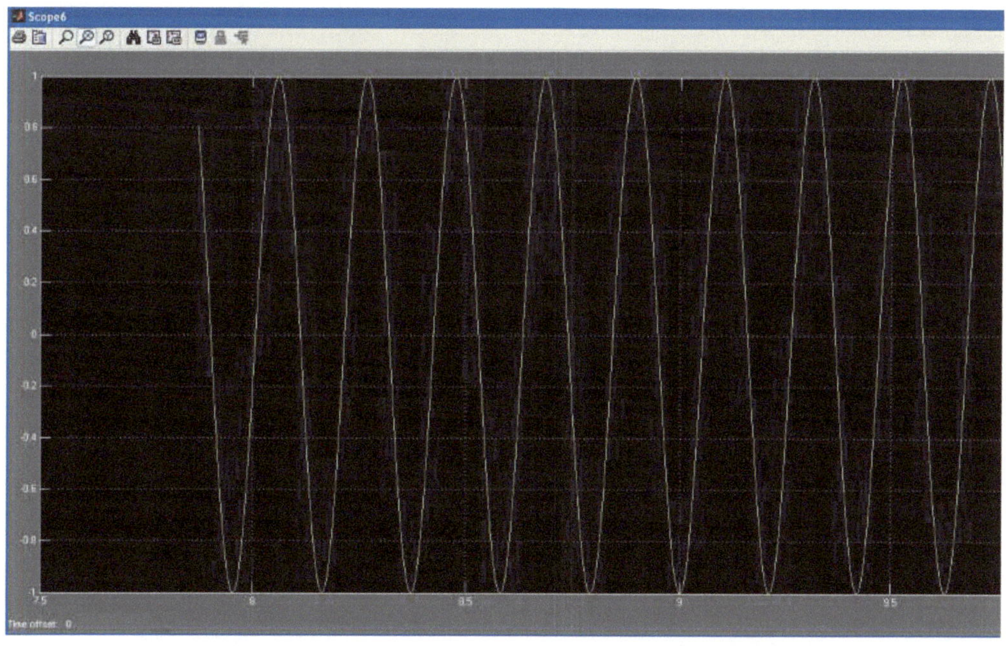

2) Phase detector output

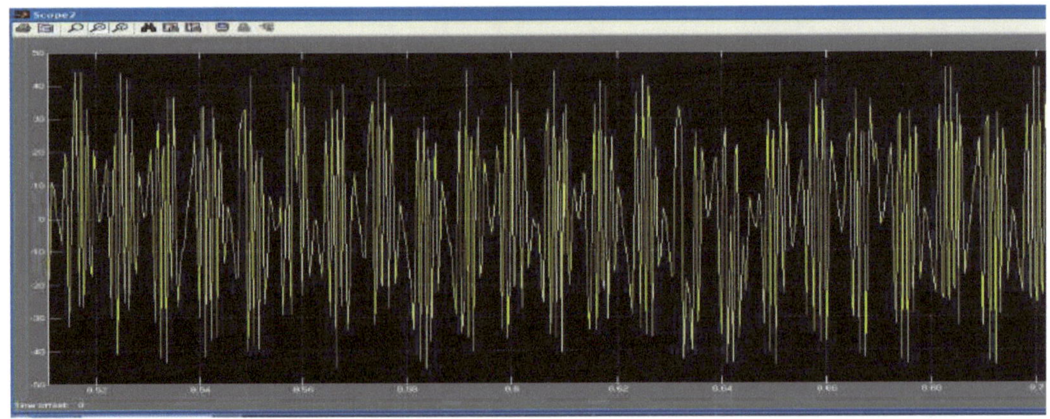

3) VCO output

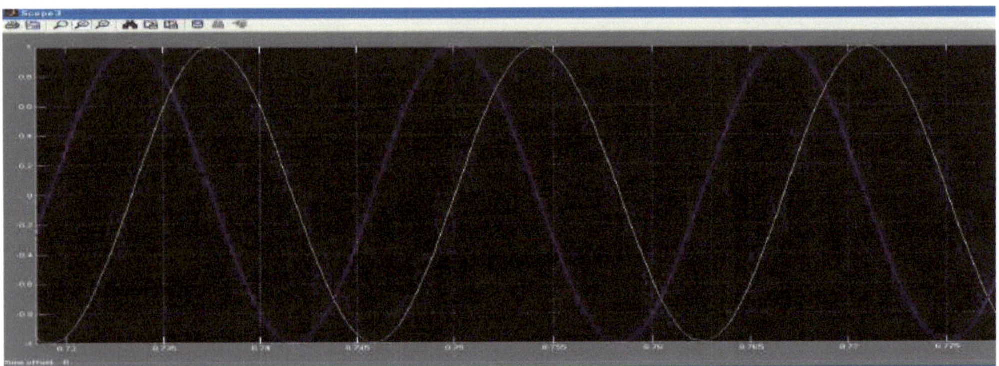

4) Loop filter output

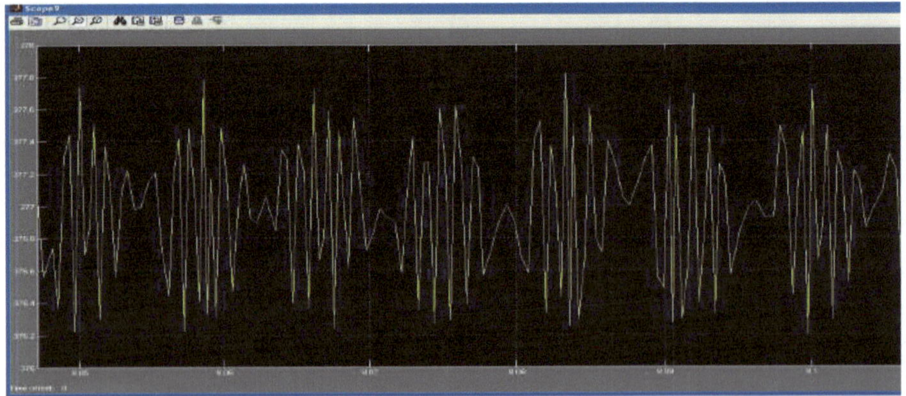

5) Schematic of the QVCO

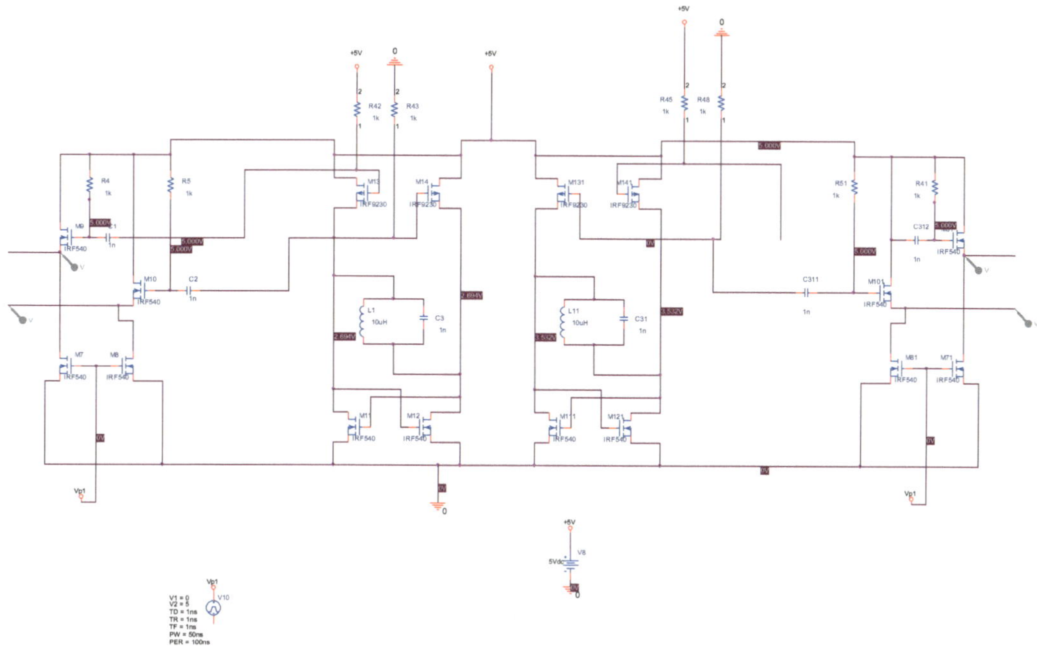

"VCO output 1 has a 2.27 Volt amplitude and a 0.2 microsecond time period (LC tank based)"

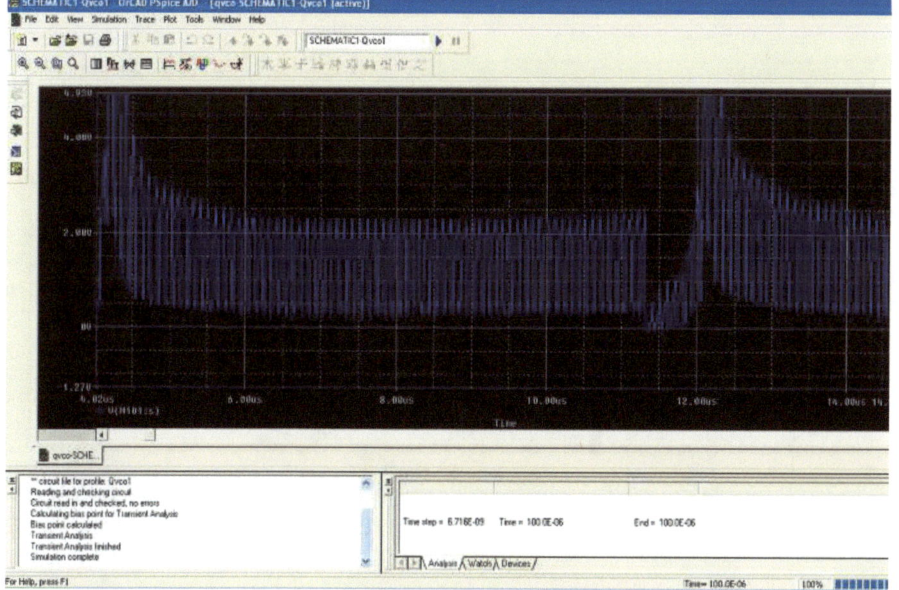

The VCO output2 has an amplitude of 2.15 volts and a time period of 0.1 microseconds (LC tank based).

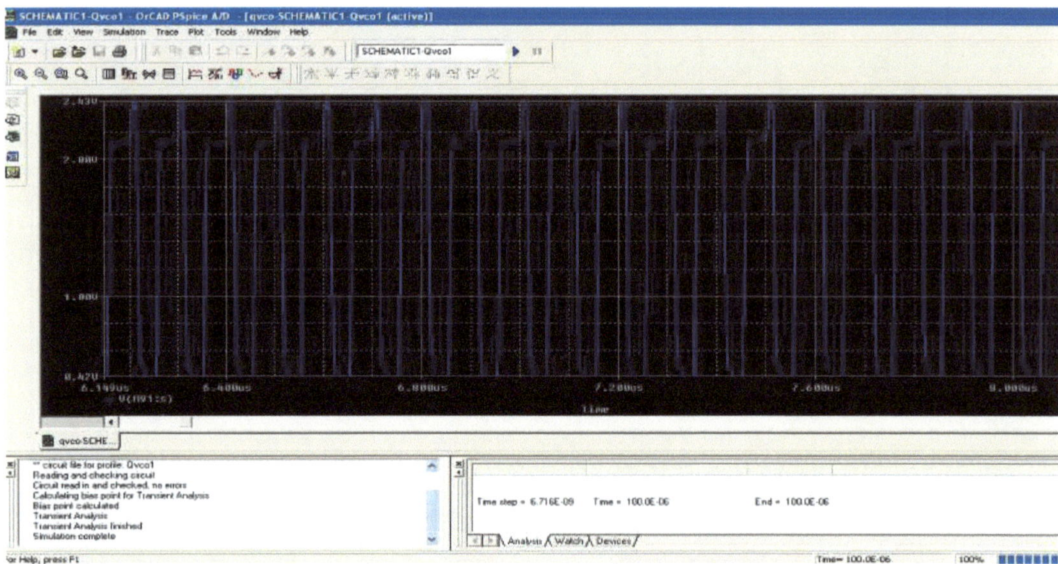

"VCO output 3 with a 2.3 volt amplitude and a 0.1 microsecond time period (LC tank based)"

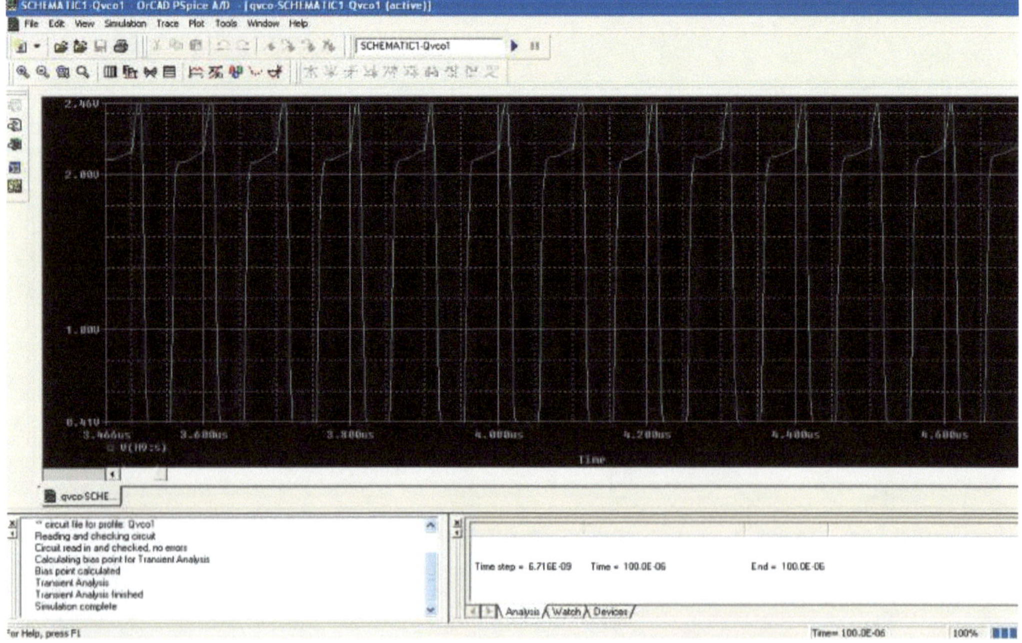

"VCO output 4 obtains amplitude 2.54 Volts and time period 0.2 microseconds (LC tank based)"

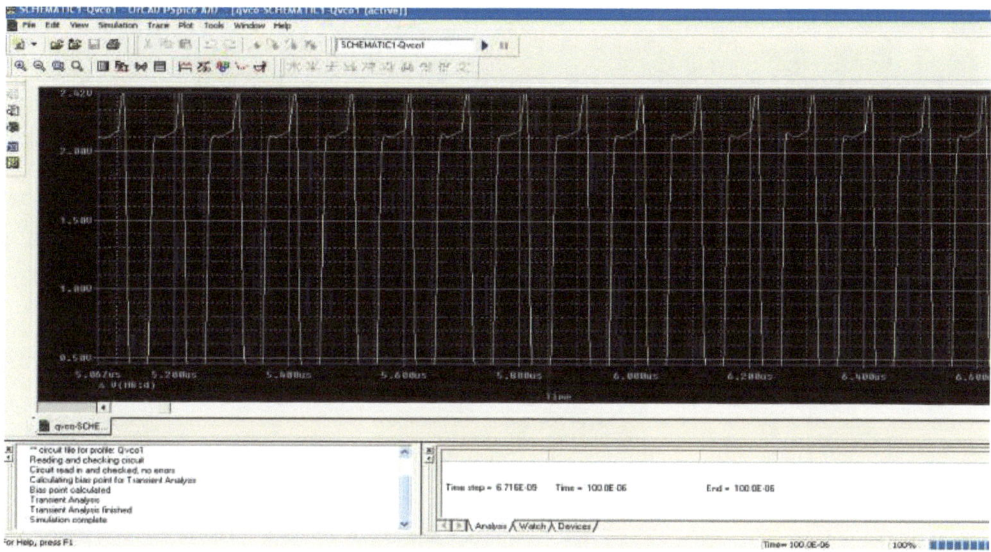

"LC Tank output with amplitude 8.83 Volts and time period 8 microseconds"

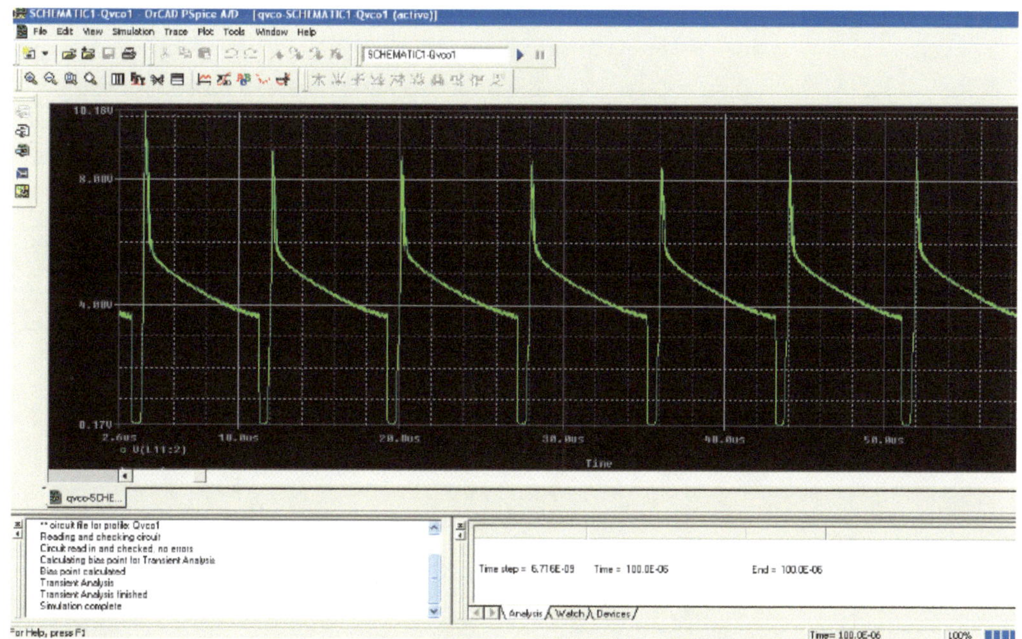

"Bias output with amplitude 4.86 Volts and time period 0.1microseconds"

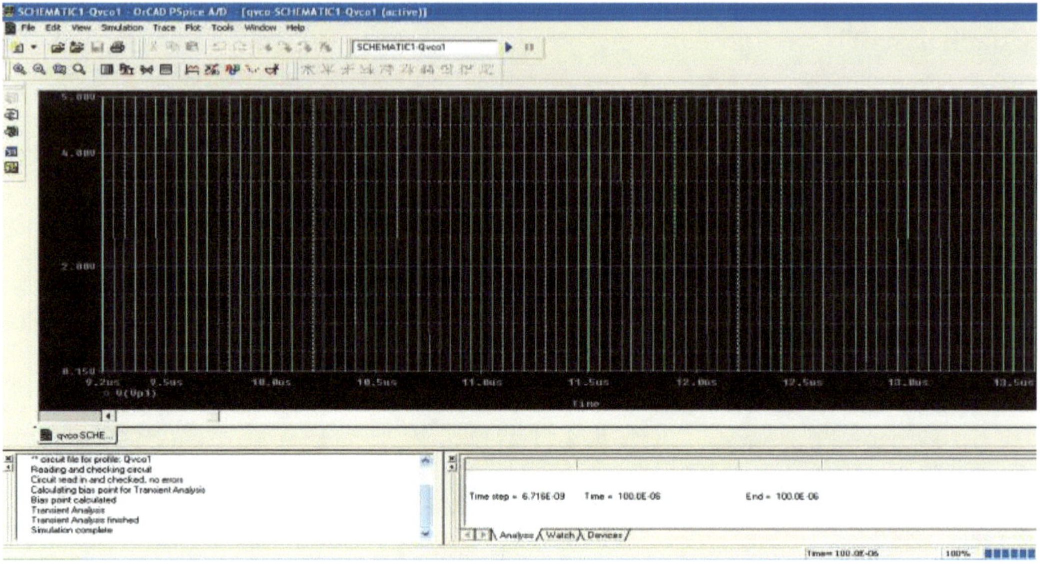

6) Schematic of double balanced Gilbert mixer

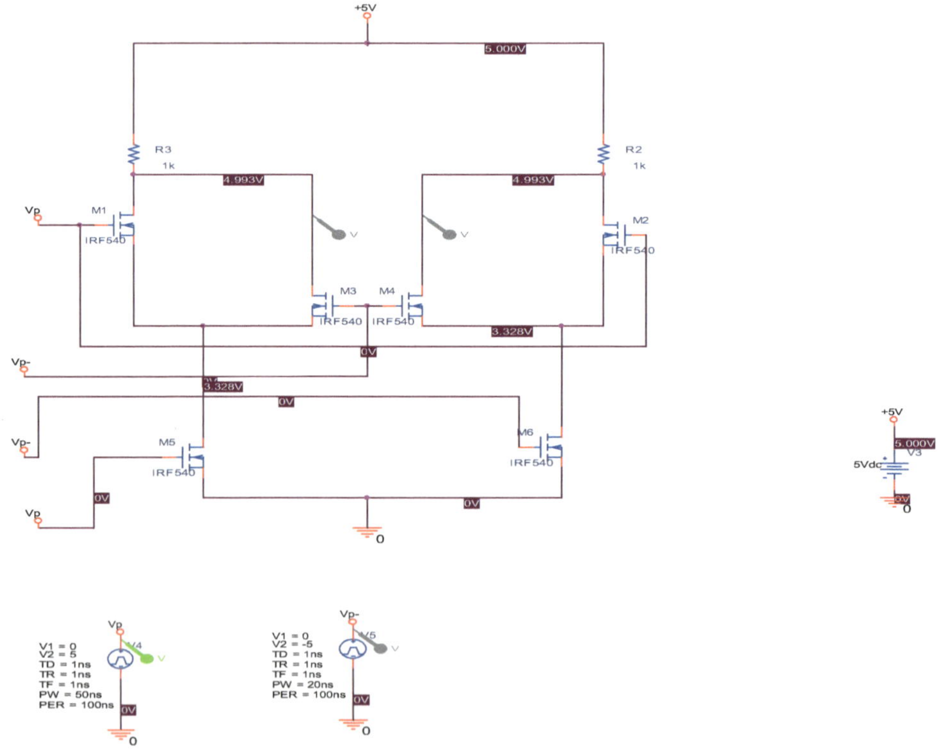

"With an amplitude of 0.854 Volts and a time period of 0.1 microseconds, the mixer is out"

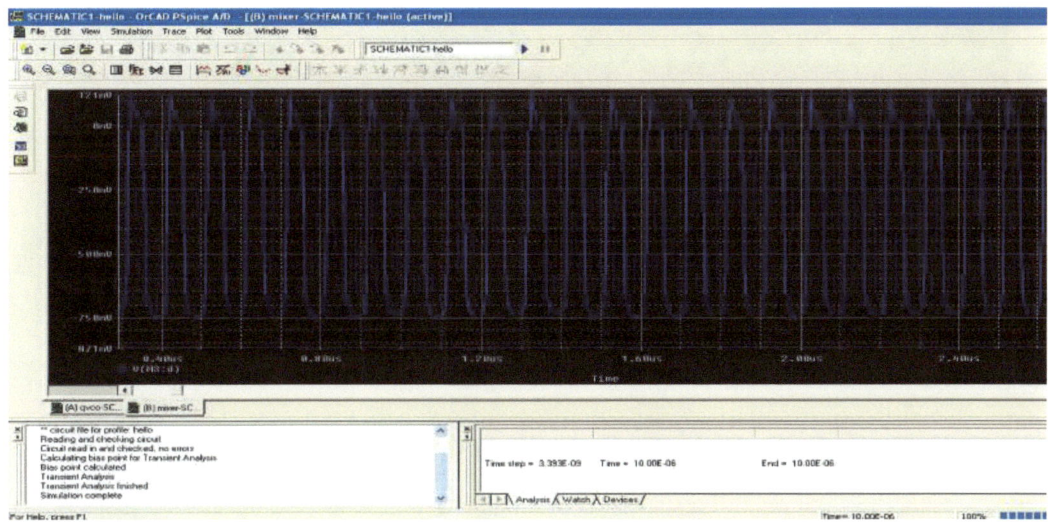

"The mixer outputs have an amplitude of 2.28 volts and a time period of 0.1 microseconds "

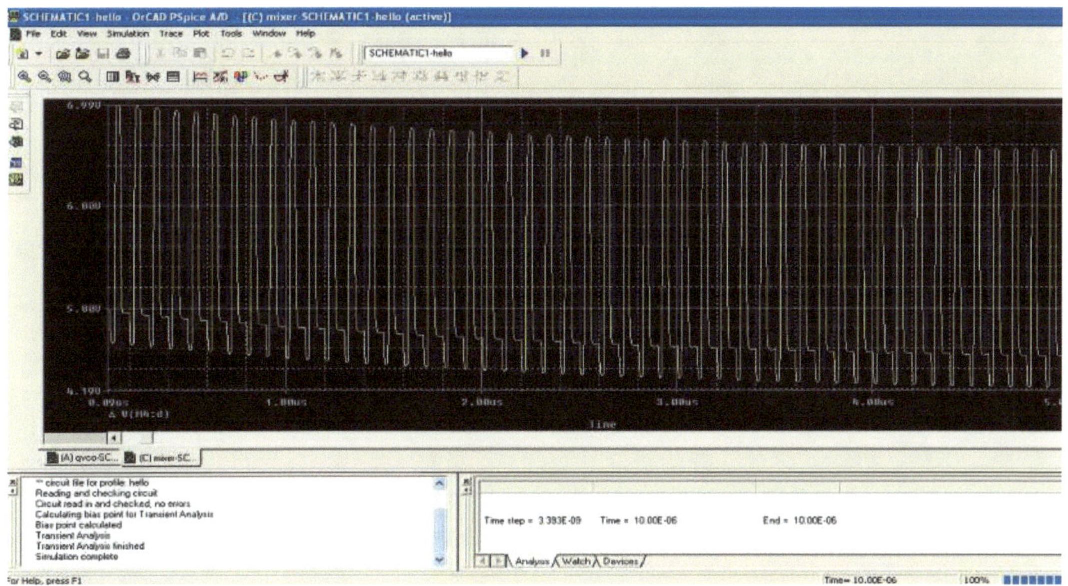

"Local oscillator output with a 0.1 microsecond period and a 5 volt amplitude"

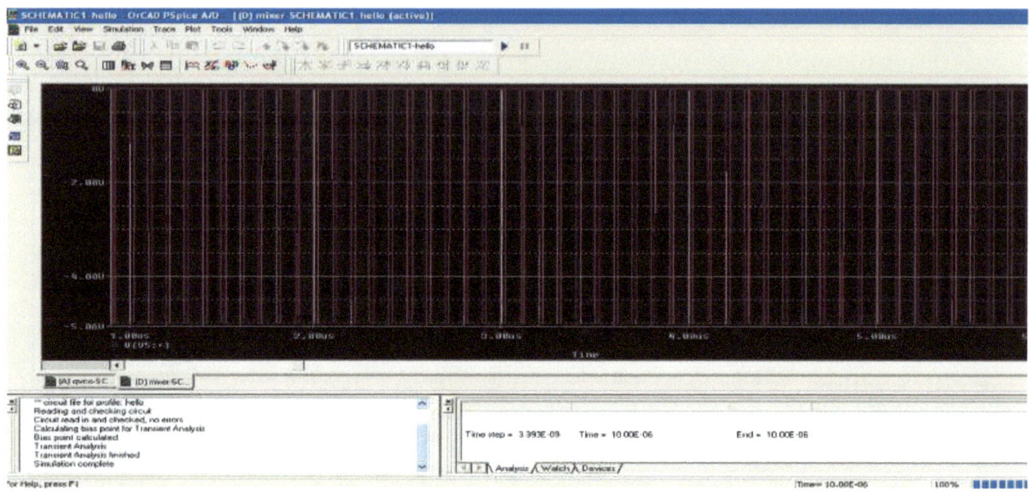

"RF output with amplitude 5 Volts and time period 0.1microseconds"

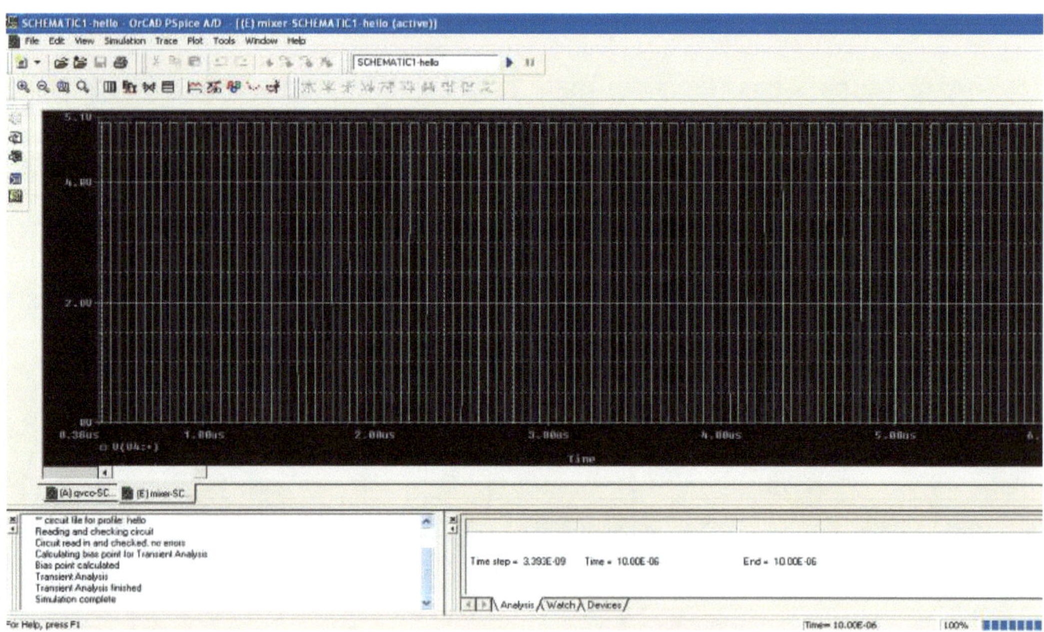

"The VCO's output amplitude is 3.4 volts, with a 15 nanosecond time period."

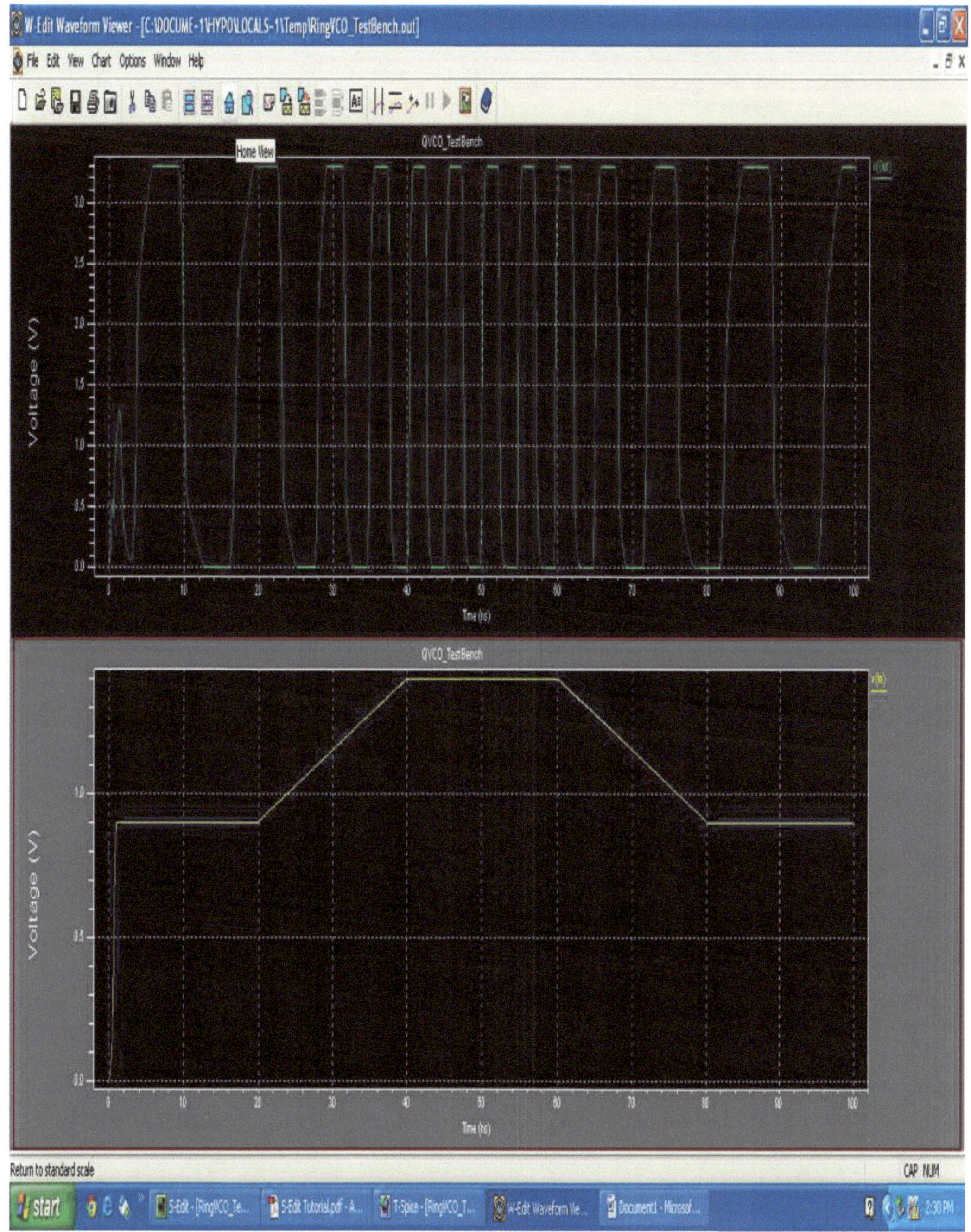

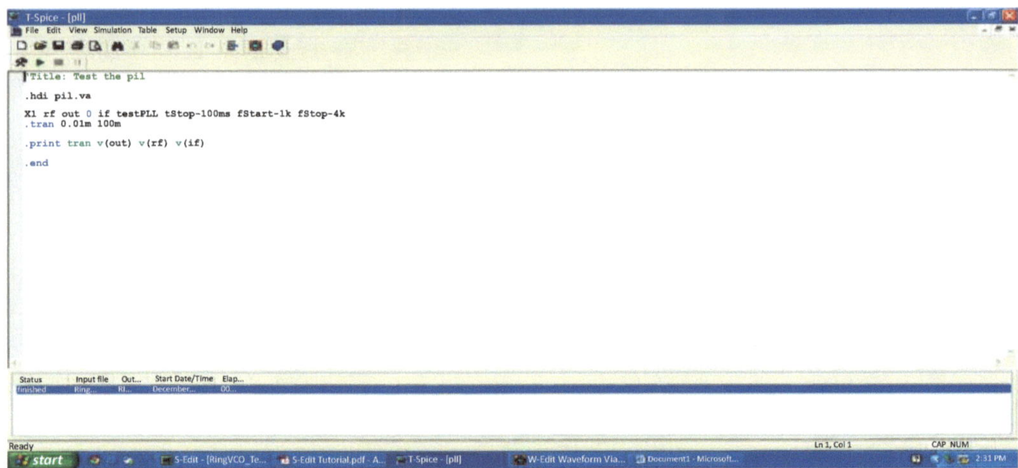

"PLL output with an amplitude of 2 volts and a period of 0.833 milliseconds"

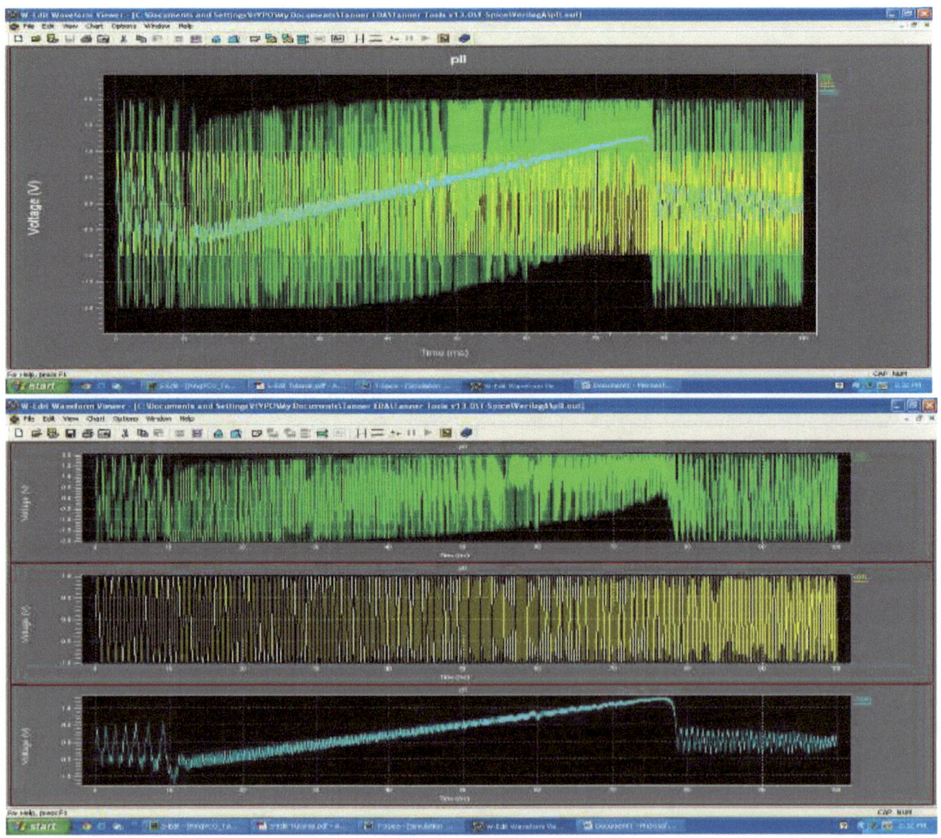

CONCLUSION

A mixed-signal broadband quadrature receiver with an integrated modem makes use of the distinctive boundaries between analogue and digital circuits to provide a high-performance IC with a small footprint and low power consumption. A multi-gigabit demodulation operation is successfully performed using a high-speed digital ASIC device combined with an analogue front-end. The suggested architecture is a fully integrated system that provides multi-gigabit modem capabilities and keeps the power budget in the sub-Watt range. The mixed-signal demodulator is capable of multi-gigabit BPSK demodulation at a data rate of 3.5 Gb/s. Multi-gigabit OOK and DBPSK demodulations are possible with the mixed-signal demodulator. We may be able to create the aforementioned system in the future using 45 nm technology. We can achieve a 12-14% area reduction system, up to 65% power reduction, and precise analysis by doing so. Area overhead may also be decreased by employing the clock gating idea in quadrature VCO.

REFERENCES

[1]   M. S. Cabric, W. Chen, D. A. Sobel, S. Wang, J. Yang, and R. W. Brodersen, "Novel radio architectures for UWB, 60GHz, and cognitive wireless systems", *EURASIP Journal on Wireless Communications and Networking.,* vol. 2006, p. 18, 2006.

[2]   A. Deparis, A. Bendjabballah, A. Boe, M. Fryziel, C. Loyez, L. Clavier, N. Rolland, and P.-A. Rolland, "Transposition of a baseband UWB signal at 60 GHz for high data rate indoor WLAN", *IEEE Microwave and Wireless Components Letters.,* vol. 15, no. 10, pp. 609-611, 2005.

[3]   A. Deparis, C. Boé, C. Loyez, L. Clavier, N. Rolland, and P.-A. Rolland, "UWB-IR transceiver for millimeter wave WLAN", *Proceedings of the 32nd Annual Conference on IEEE Industrial Electronics (IECON '06),* 2006pp. 4785-4789 Paris, France

[4]   N. Devulder, A. Deparis, and I. Telliez, "60?GHz UWB transmitter for use in WLAN communication", *Proceedings of the International Symposium on Signals, Systems and Electronics (ISSSE '07),* 2007pp. 371-374 Montreal, Canada

[5]   Park Ji-Yong, Jeon Seong-Sik, Wang Yuanxun, and T. Itoh, "Integrated antenna with direct conversion circuitry for broad-band millimeter-wave communications", *IEEE Trans. Microw. Theory Tech.,* vol. 51, no. 5, pp. 1482-1488, 2003.
[http://dx.doi.org/10.1109/TMTT.2003.810128]

[6]   P. Smulders, "Exploiting the 60 GHz band for local wireless multimedia access: Prospects and future directions", *IEEE Commun. Mag.,* vol. 40, no. 1, pp. 140-147, 2002.
[http://dx.doi.org/10.1109/35.978061]

CHAPTER 12

Pre Placement 3D Floor planning of 3D Modules Using Vertical Constraints For 3D IC'S

J. Zahariya Gabrie[1,*], **Ravi R.**[1], **R. Kabilan**[2] and **M. Philip Austin**[2]

[1] *Department of CSE, Francis Xavier Engineering College, Affiliated with Anna University, 103/G2, Bypass Road, Vannarpettai, Tirunelveli, Tamil Nadu 627003, India*

[2] *Department of Electronics and Communication Engineering, Francis Xavier Engineering College, Tirunelveli, India*

Abstract: This project focuses on wire length reduction throughout the 3D floor layout stage. The 3D cell layout stage is part of the floor planning process. Previously, it was expected that the entire module would be placed on a single device layer. They don't consider how a module's cells may be dispersed across many device levels to reduce the cable length. Each of the device layers is assigned to one of the cells that make up a module (a 2D module is converted into a 3D module). To place cells in three dimensions, several constraints are used. The placement aware constraints are a set of constraints that determine whether a 2D module may be turned into a 3D module. The vertical alignment of identical submodules owing to the same planar placement requirement is referred to as vertical constraint. The size of the solution will be reduced as a result of this. A 3D floor design module packing method is proposed by the author. Calculating the wire length and taking into consideration the feasibility requirement, a smaller solution area is used to arrange the 3D cells in an initial set of floor layouts. After finding the best floor design, the modules are packed using a packing algorithm, and the technique is finished. A placement aware 3D floor design method is the name of the approach, which is developed in C++ and operates on Fedora Linux.

Keywords: AGC, CMOS, Error detector, VCO,PLL.

INTRODUCTION

VLSI refers to the process of combining thousands of transistors onto a single chip. VLSI arose in the 1970s while complex semiconductor and communication technologies were being developed. The microprocessor is a VLSI device.

* **Corresponding author J. Zahariya Gabrie:** Department of CSE., Francis Xavier Engineering College, Affiliated with Anna University, 103/G2, Bypass Road, Vannarpettai, Tirunelveli, Tamil Nadu 627003, India; E-mail: zahagabs@gmail.com

S. Kannadhasan, R. Nagarajan, Alagar Karthick, K.K. Saravanan & Kaushik Pal (Eds.)

Speed, power, and space are all important factors in VLSI (layout). Signal propagation delays through gates and wires occur even for areas just a few micrometres wide. The operating speed is so rapid that when the delays build, they resemble the clock speeds. High operational frequencies can result in increased power usage. This has two effects: electronics absorb battery energy faster, and heat dissipation increases. When heat is paired with the fact that surface areas have reduced, the circuit's integrity is jeopardised.

Circuit component layout is a profession that is identical to anything in the electronics area. The fact that we have different possibilities for doing so makes our situation unique: several layers of diverse materials on the same silicon, various arrangements of smaller components for the same component, and so on. There is a trade-off in the circuit between power dissipation and speed; if one is maximised, the other suffers. The bulk of today's VLSI designs are classified as analogue, ASIC (Application Specific Integrated Circuits), or SoC (System-on-Chip) designs (System on Chip). The purpose of research and development in this discipline is to optimise area, speed, and power. Alternative strategies for generating improved system-level performance are becoming more significant as a result of the growing difficulties in sustaining technological achievements *via* conventional scaling at a rate corresponding with Moore's law. 3D technology has the potential to provide significant performance increases over 2D for multiple generations. In 3D physical design, planning, buffering, and temperature control are all key issues.

Alternative strategies for generating improved system level performance are becoming more significant as a result of the growing difficulties in sustaining technological achievements *via* conventional scaling at a rate corresponding with Moore's law. 3D technology has the ability to significantly increase performance across numerous generations. 3D technology refers to the production of a single integrated circuit whose functional components (for example, transistors) extend in three dimensions, *i.e.*, the vertical arrangement of several bare integrated circuits in a single package. Because of the vertical extension of chips in 3D technology, improved performance and power efficiency are achieved *via* the use of chip area, decreased connection length, and increased packing density.

Transitioning from conventional single chip packaging to 3D technology saves a lot of space and weight. The key limiting factors for 3D technology are connection capacity, thermal characteristics, and the required durability. 3D technology has been claimed to provide a 40-50 times decrease in size and weight when compared to conventional processes. All of these reductions are the result of eliminating the overhead weight and size that conventional technology entails.

One of the most critical considerations in packaging technology is the chip footprint, or the area of the printed circuit board occupied by the chip. "Silicon efficiency" refers to the ratio of total silicon footprint area to substrate area. As a consequence, the silicon efficiency in any 2D technology can never exceed 100%; nevertheless, owing to many overlapping footprints within one stack in 3D technology, this limit is broken. As a consequence, shortening the stack's connections minimises the propagation latency between chips.

RELATED WORKS

Through-silicon *Via* Planning in 3-d Floorplanning

None of the studies take into account TSV's size and location. Previously, they were used as points while calculating the wirelength. For wire length reduction, the position and area of TSVs are taken into account. Because a significant mistake in the wirelength estimate reduces the optimality of the floor plan results, accurate results may be obtained by considering the area and location of TSVs. Previously, the wirelength was estimated using half the perimeter length of the bounding box holding the pins without taking into account the position of TSVs.

Without considering the position of TSVs, the bounding box is tiny; however, when the position of TSVs is taken into account, the bounding box becomes large, and the bounding box's half perimeter length grows. As a result, a more accurate wire length estimation is attained.

At first, probabilistic TSV planning was chosen at high temperatures, and as the temperature dropped, thorough TSV planning was preferred. As a result, the heat issue was also brought up. By 57 percent, our technique outperformed the postprocessing TSV planning algorithm.

Efficient Thermal *via* Planning Approach and its Application in 3-d Floorplanning

During three-dimensional (3-D) floorplanning, the 3D floorplanning method has been devised to account for thermal *via* (T-via) planning. It also incorporates dynamic TVP into the three-dimensional floorplanning process. A series of simplified interlayer and interlayer TVP sub problems are used to solve the temperature-limited TVP issue. Each subproblem is recast as a convex programming problem, and the best solution for the detailed T-via distribution is found. A two-stage strategy is used to construct an integrated TVP and 3-D floorplanning algorithm based on the TVP solution.

The actual 3-D representation, which has too much duplication in adding data structure on the z-axis and is inefficient in both time and space, is one of the two types of 3D floorplan representation. Next, due to the device layer structures, 2-D-array-based representations have a substantially larger solution space, making the 3-D floorplanning issue much more complicated and resulting in longer running time and/or worse solution quality.

Rectangular 3d Wirelength Distribution Models

Three current 3D wirelength distribution models, originally created for square modules, are expanded to handle rectangle 3D blocks in this paper. From the 3D chip level to the 3D block level, the expanded models enable stochastic wirelength prediction. These newly created rectangular models have been subjected to comparative and qualitative research. The experimental findings show how these models are compared in terms of efficiency and computing time. The suggested approach will be tremendously valuable in the interconnect-centric 3D floorplanning for quickly estimating wirelength.

Wire-length Distribution of Three-dimensional Integrated Circuits

The wire length distribution of three-dimensional integrated circuits is calculated using the same approaches as the wire length distribution of two-dimensional integrated circuits. The 3D distribution model is derived from the 2D distribution model. The wire length distribution of 2D integrated circuits is:

$$f_{2D}(l) = \Gamma M_{2D}(l) I_{2D}(l) \tag{1}$$

Where Γ is a normalization constant, $M_{2D}(l)$ is the number of gate pairs separated by length l and $I_{2D}(l)$ is the number of interconnections between logic gate pairs. For 3D IC,

$$M_{3D}(l) = M_{3D\text{-}inter}(l) + M_{3D\text{-}intra}(l) \tag{2}$$

EXISTING METHOD

Slicing and nonslicing floor plans are the two types of 2-D floor planning. It has a smaller solution space, which means it runs quicker. Unfortunately, the majority of real-world designs are nonslicing, necessitating the use of complicated topological representations for floor planning. These representations may be identified based on the solution space's P-admissibility [13], which meets the following criteria: 1) the solution space is limited; 2) all solutions are practical.

Between their topological representations and the accompanying floor plans, both SP and TCG have a one-to-one mapping. BSG, on the other hand, creates many representations for a single packing, implying a broader solution space than SP and TCG. The true-3-D and quasi-3-D representations of 3-D floor planning may be divided into two categories. True-3-D representations are created by extending a 2-D representation to a genuine 3-D structure.

For the perturbation of the solution space, these representations are used to execute intralayer operations and interlayer operations across distinct levels. For 3-D microprocessors, Hung *et al.* introduced an interconnect and temperature aware floor planning technique. During optimization, the programme evaluates the power usage in cables. For wirelength optimization, Li *et al.* suggested a hierarchical 3-D floor planning technique. A scalable temperature and leakage sensitive 3-D floor planning method was recently proposed by Zhou et al.

PROPOSED METHOD

Pre Placement 3d Floor Planning of 3d Modules Using Vertical Constraints for 3d Ics

During the design phase of this project, three issues were discussed: wirelength, area, and runtime reduction. A single layer is examined in a conventional floor plan, where the modules/blocks in that single layer are adjusted to minimise wirelength, runtime, and area. Multiple 2D layers are piled vertically in a 3D floor plan to create a 3D structure, and a 3D floor plan of 2D modules is created. When 2D modules are converted into 3D modules by separating them into sub-modules and assigning these sub-modules to separate device layers, the overall wirelength is drastically reduced. In other words, 3D cell placements inside particular modules may enhance floor planning methods. This cuts down the amount of time it takes to complete a task.

Place aware constraints, also known as vertical constraints, are used to make the choice to convert a 2D module into a 3D module. The requirements are: 1) each submodule must have the same dimensions; and 2) each submodule must have the same number of submodule 2) Sub-modules are stacked in device layers one after the other. 3) All submodules are on the same plane. 4) There can only be one split module submodule per device tier. Each solution is a parent that creates children while conducting reproduction. It's done to figure out what a 2D module's total wirelength is and how it will alter when it's partitioned. 2D and 3D wirelength distribution models with a rectangular shape. Swap, invert, insert, rotate, and exchange procedures are used to modify each floor plan. The optimum solution is determined by calculating the fitness for each solution and determining which solution has the best fitness is shown in Fig. (**1**).

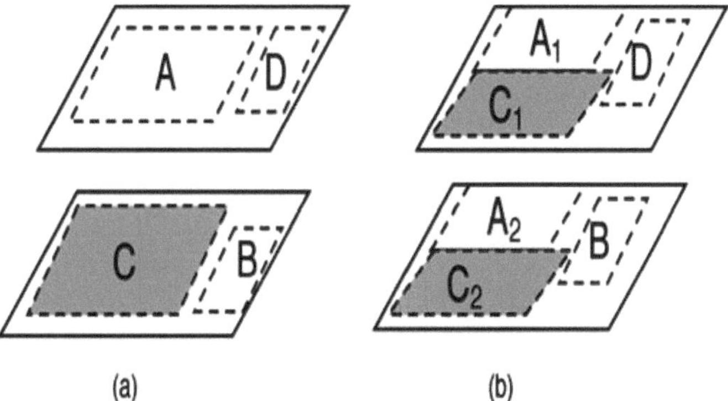

Fig. (1). (**a**) shows a 3D layout of 2D modules. (**b**) shows a 3D layout of 3D modules.

Vertical Constraints on Sequence Pairs

Sequence pair(SP) Representation

As a result, a sequence pair offers relative module locations rather than physical information. They're known as a clustered sequence pair (GSP).

Two Layer Feasibility Condition

Vertically aligned submodules are identified with the same letter and differentiated by subscripts indicating their device levels. Please note that A1 and B1 are only limited along the +X axis in layer 1 is shown in Fig. (**2**).

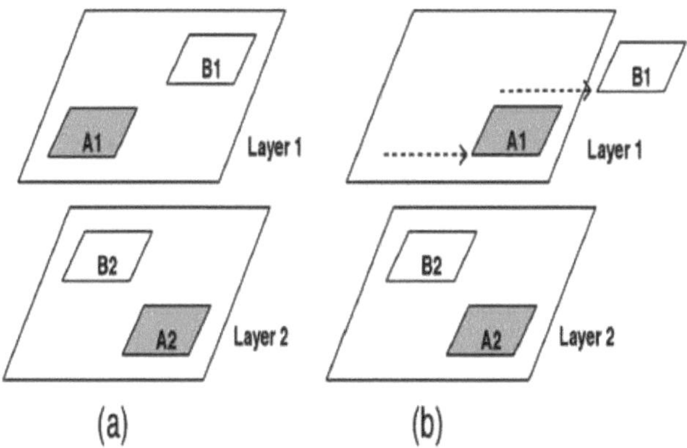

Fig. (2). (a,b) Vertically constrained module pairing.

Module Packing with Vertical Constraints

Module Packing by LCS and Lateral Shifting (LCSLS)

The LCSLS is a fast module packing technique with vertical restrictions. Because the same method is true for alignment along the -axis, we address the packing with limited module alignment along the -axis is shown in Fig. (**3**).

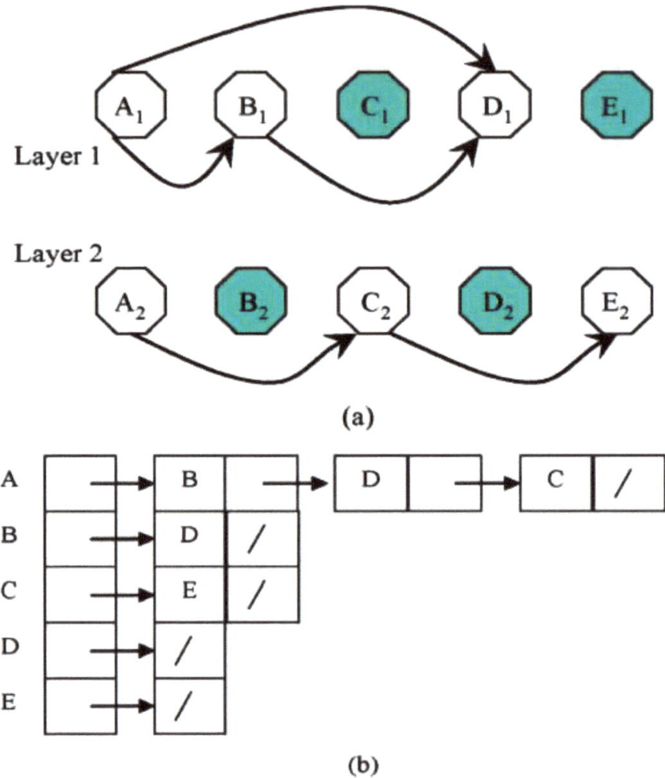

Fig. (3). A restricted adjacency list.

By inserting edges between submodules, a tiny sub-graph is produced in each device layer. By substituting each submodule with its group name, we may unite these subgraphs.

Placement Aware 3-d Floor Planning Algorithm

Design Flow of the 3-D Floor Planning Algorithm

We begin with a limited number of randomly generated floorplans that do not contain any split modules. By statistically assessing the reduction in wirelength during disturbance, we investigate the potential of 3-D placement inside a module (module-split move). At runtime, if a module is divided into many pieces, its submodules are subjected to vertical limitations. Furthermore, we limit movements to those that meet the feasibility criteria is shown in Fig. (**4**).

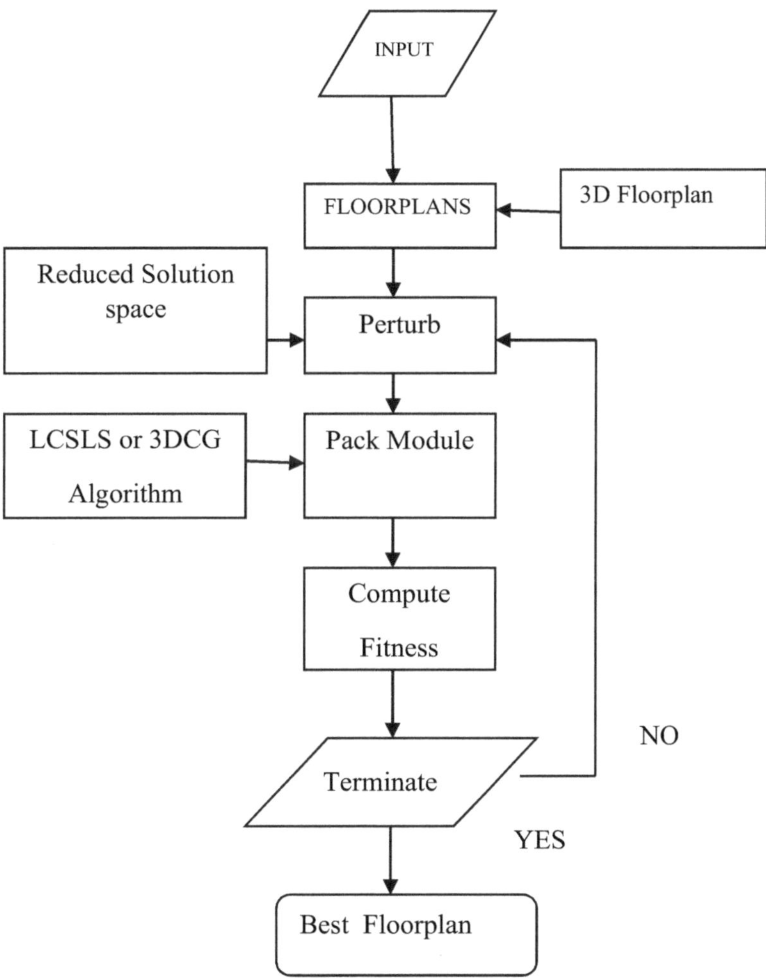

Fig. (4). Flow chart of placement aware 3D floor planning.

Stochastic Combinatorial Optimization

Each person in the population is a one-of-a-kind solution. Parents are subjected to a random "reproduction" process that results in children.

Cost Estimation of Placement Inside 3-D Modules

As a result, the total wirelength within a 2-D module, as well when how it will change as the module is divided into submodules, is unknown. Statistical wirelength prediction methods are helpful in this situation.

RESULTS AND DISCUSSION

The result of the C++ algorithm, which was executed on the Fedora Linux operating system, is shown below (Fig. **5**).

Fig. (5). Input Phase.

This is the beginning of the input phase. As an input, the number of layers, modules, sub-modules, I/O channel, and three dimensions of the initial layer are provided is shown in Fig. (**6**).

The population of the original set of floor plans is shown below. The three dimensions of all modules are computed when the user inputs are received. Then, as illustrated above, three different types of floor plans are filled. Splitting the modules into sub-modules and arranging the sub-modules in various levels to fulfil the vertical limitations yield the best floor plan of the three.

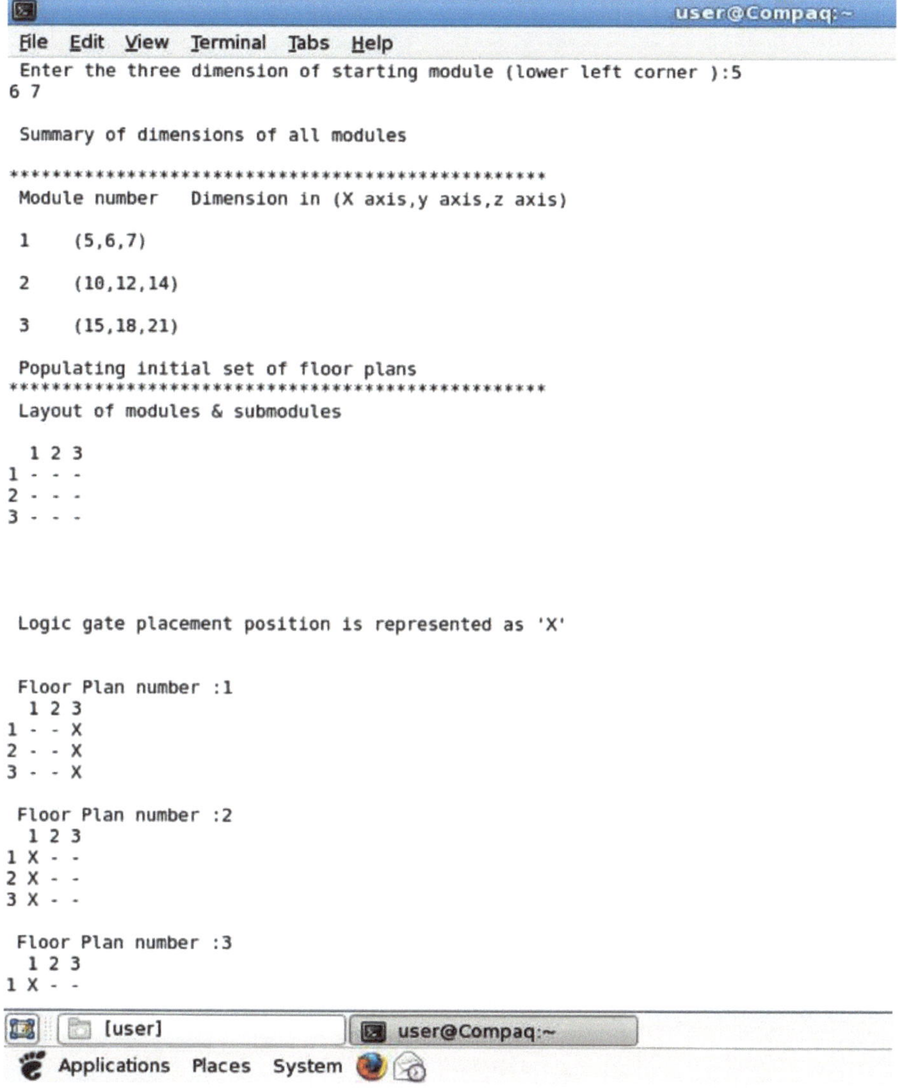

Fig. (6). Floor plan populating phase.

The processing step is shown, and floorplan 1 has been processed. Each floorplan is handled by calculating the wirelength, maintaining feasibility requirements, and packing modules using a module packing algorithm. The module packing is carried out in order to determine the module co-ordinates in the processed floorplan. Calculating keff, Neff, and TSV yields the efficacy of each layout is shown in Fig. (7).

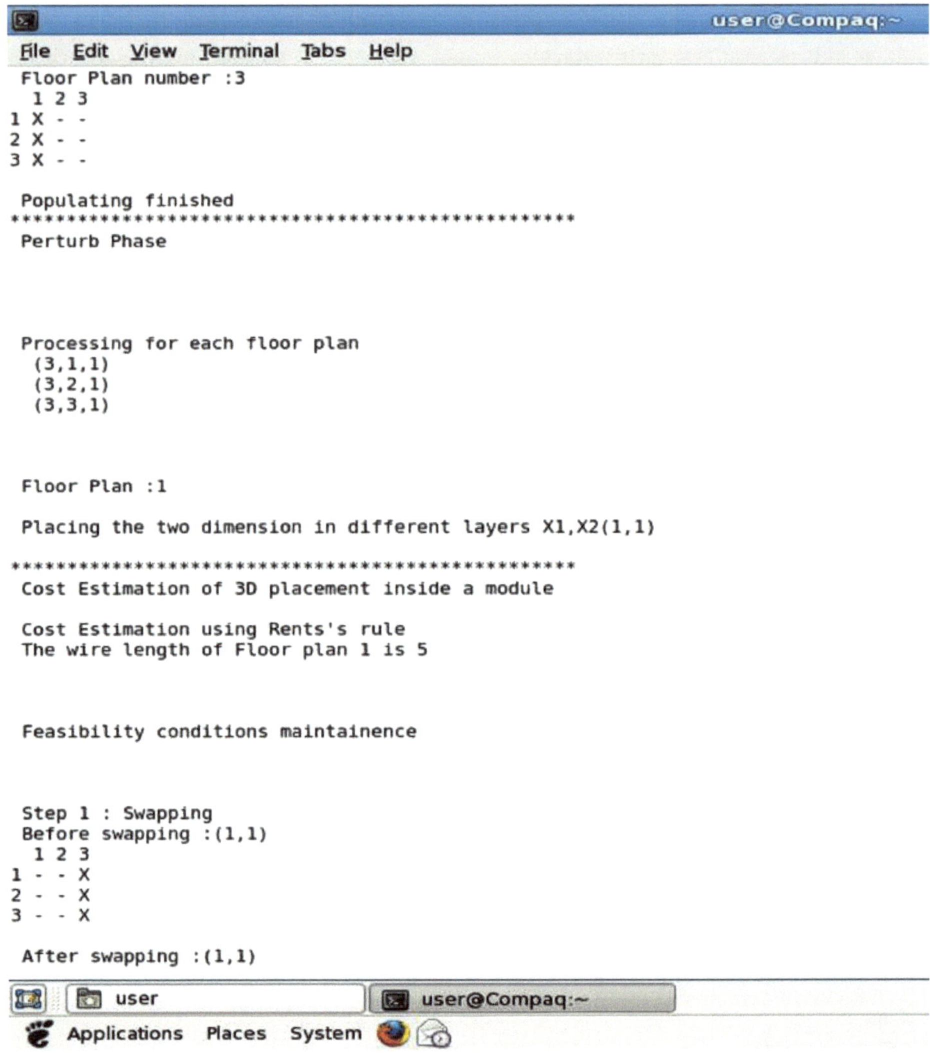

Fig. (7). Processing phase of floorplan 1.

This illustrates how floorplan 2 was processed is shown in Fig. **(8)**.

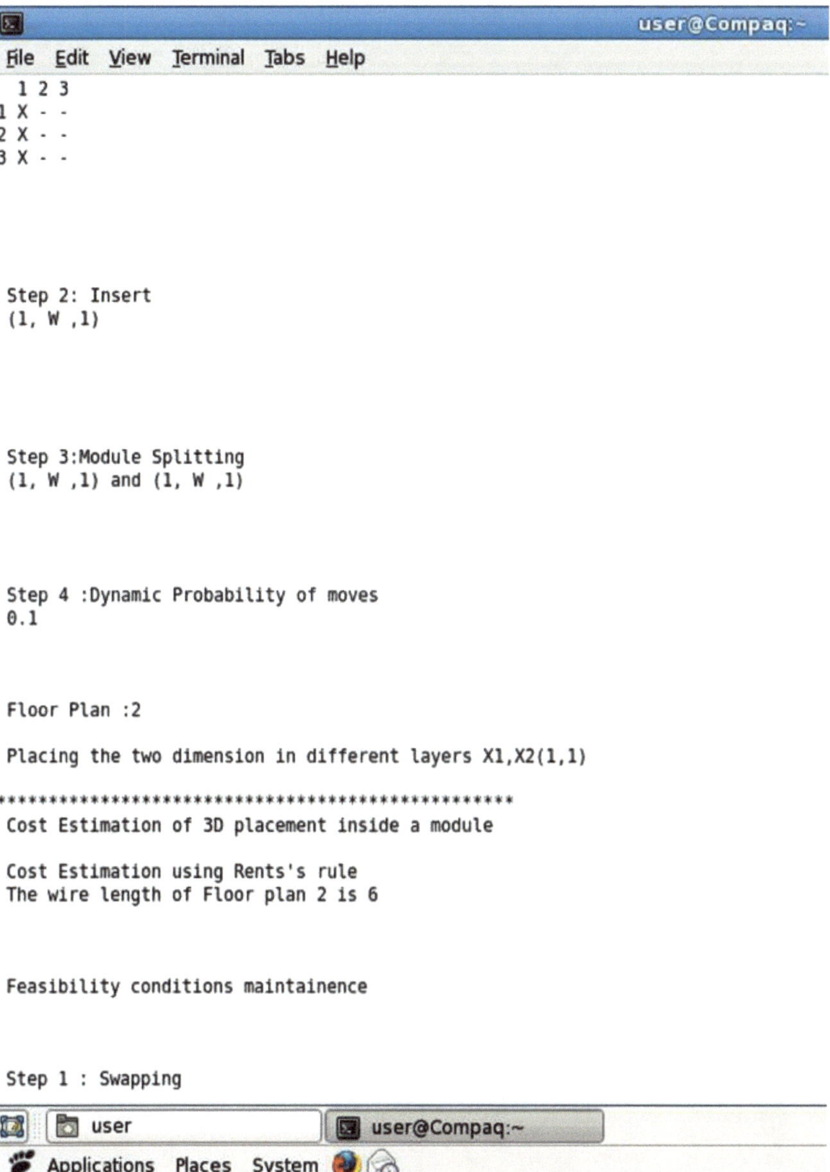

Fig. (8). Floorplan 2 processing.

This diagram shows how floorplan 3 was handled. Fig. (**9**) Floorplan 3 processing continued.

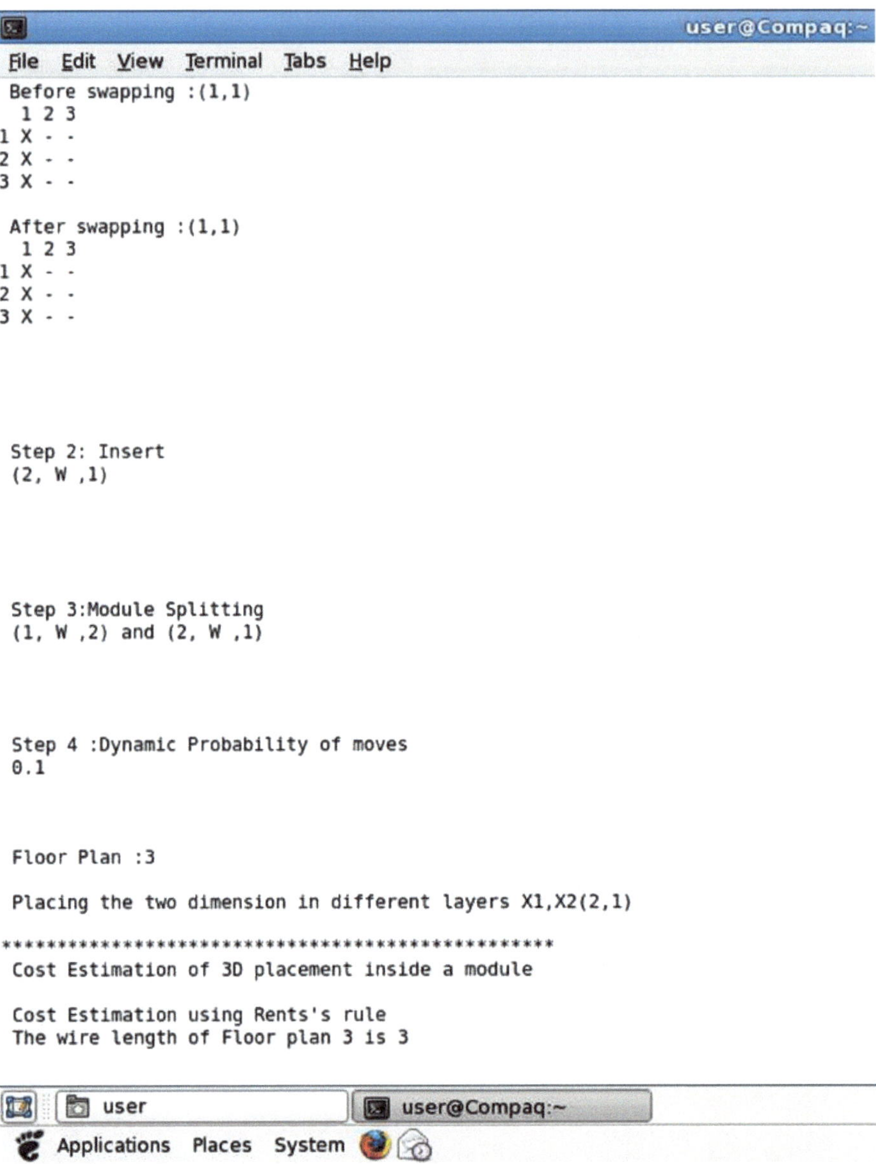

Fig. (9). Floorplan 3 processing.

When compared to other floor plans, the wire length of floor plan 3 is less, which is why it was chosen as the best layout is shown in Figs. (**10-12**).

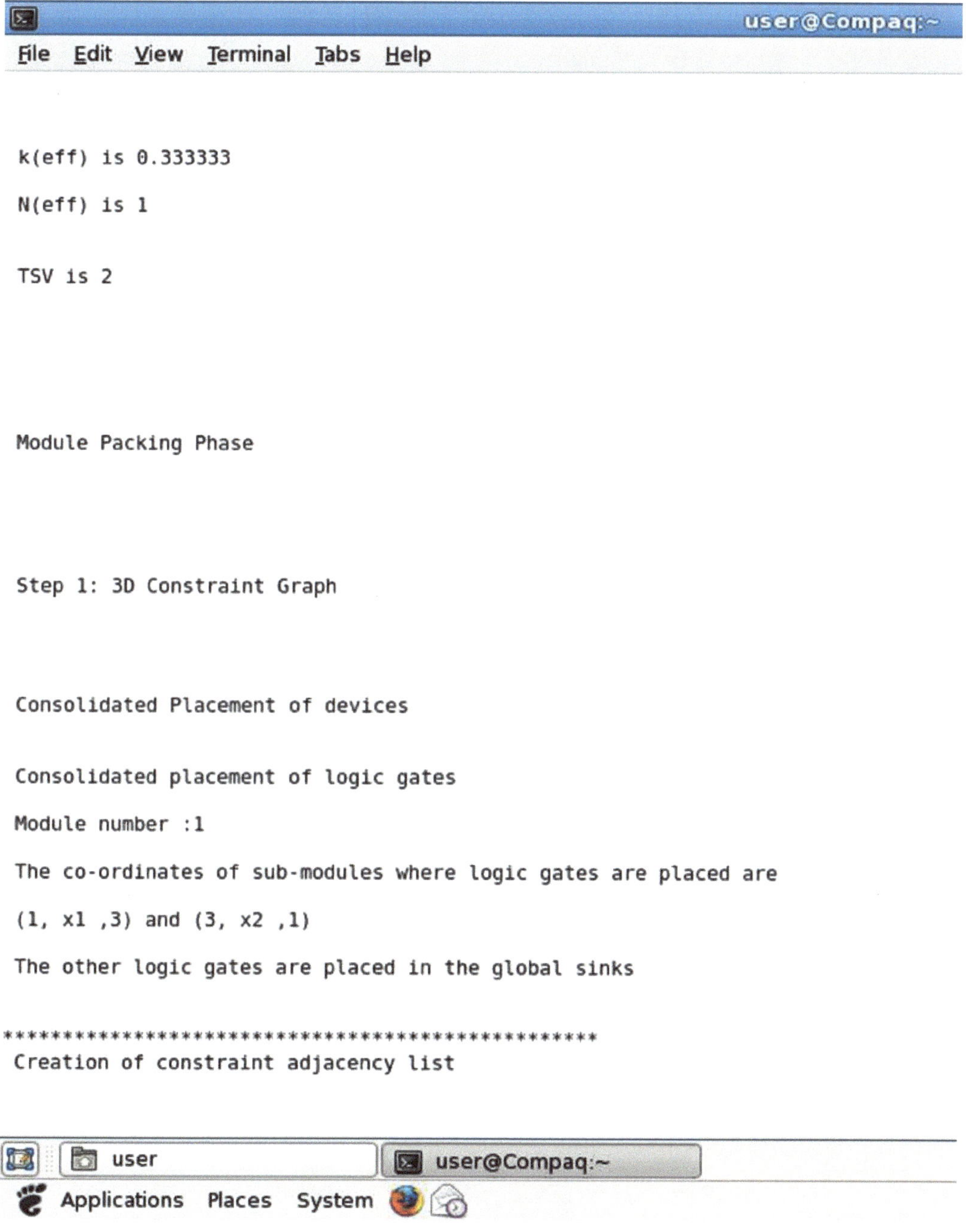

Fig. (10). Phase of module packing.

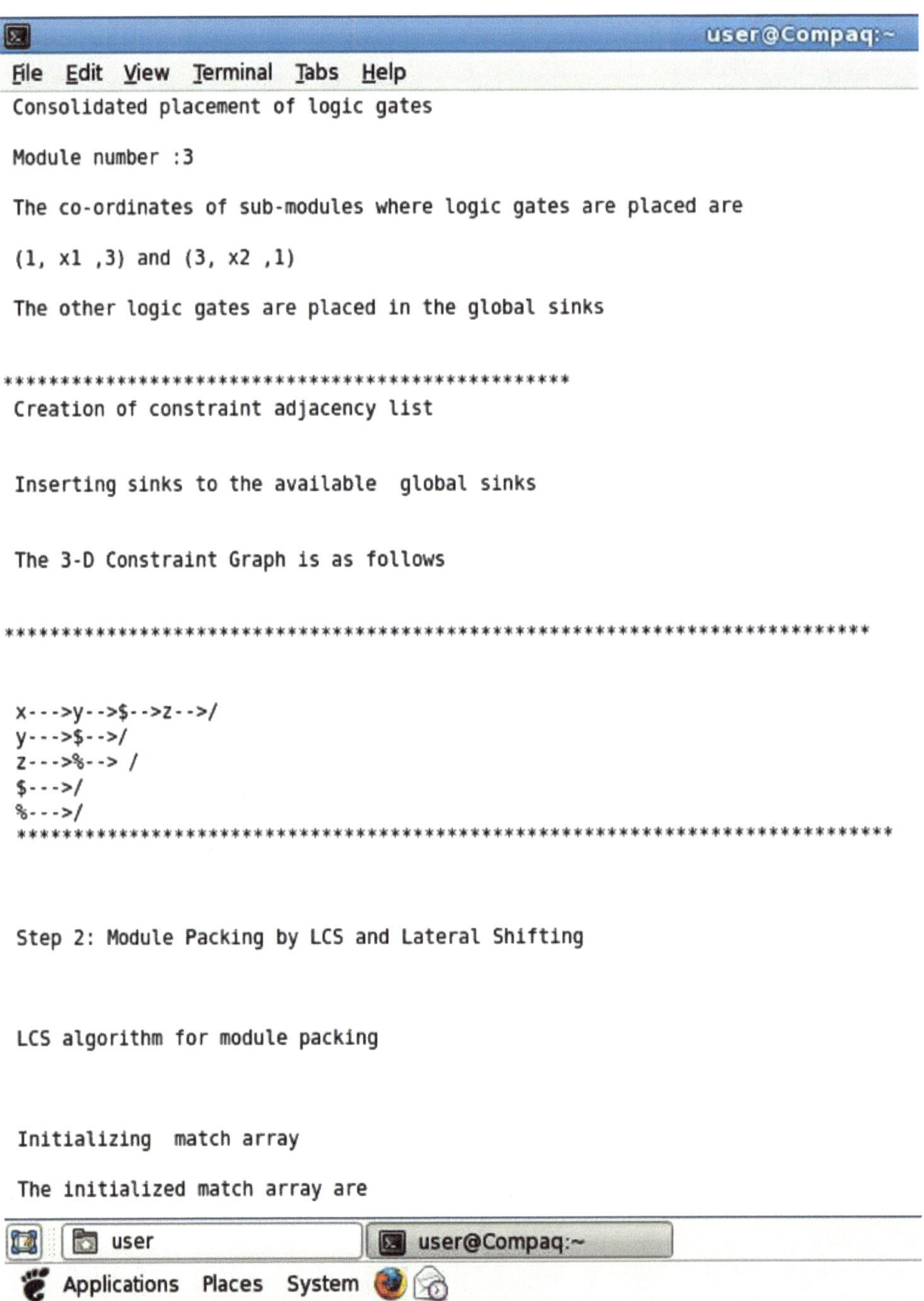

Fig. (11). Module packing phase continued.

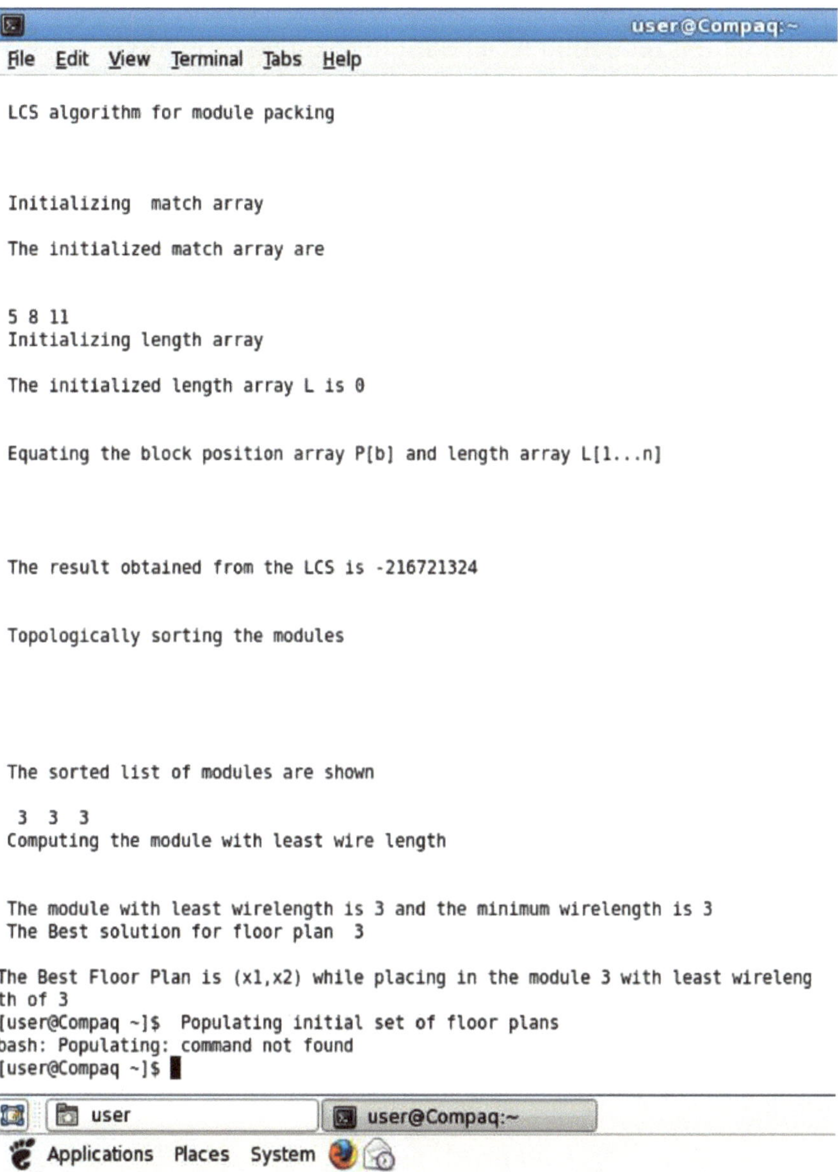

Fig. (12). Best floor plan analysis.

CONCLUSION

The vertical limitations and placement of logic gates inside the modules are taken into account by the placement aware 3D floor planning method. It investigates the idea of transforming a 2D module into a 3D module by breaking it into numerous sections, vertically aligning them using vertical constraints, and statistically calculating the wiring reduction as a result of the transformation. A set of vertical constraint feasibility conditions will also be created to specify the relationships between sequence pairs representing different device levels. The vertical limitations are also preserved when using the module packing procedure to obtain the locations of the modules.

The placement of 3D cells is done from the original set of floorplans by calculating the wire length and considering the feasibility requirement for decreased solution space. The modules are then packed using the packing algorithm, and the method is completed after the optimum floorplan has been obtained. The suggested placement aware 3D floorplanning algorithm outperforms the CBA and 3D – STAF algorithms in terms of wire length reduction.

REFERENCES

[1]   M. -C. Tsai, T. -C. Wang, and T. Hwang, "Through-silicon via planning in 3-D floorplanning", *IEEE Transactions on Very Large Scale Integration (VLSI) Systems.*, vol. 19, no. 8, pp. 1448-1457, 2011.

[2]   X. He, S. Dong, Y. Man, and X. Hong, "Simultaneous buffer and interlayer *via* planning for 3D floorplanning", *10th International Symposium on Quality Electronic Design.*, 2009 pp. 740-745 San Jose, CA, USA.

[3]   Sandro Sawicki, Gustavo Wilke, Marcelo Johann, and Ricardo Reis, "A cells and I/O Pins partitioning refinement algorithm for 3D VLSI circuits", *CLEI Electr. J.*, vol. 13, no. 3, p. 1, 2010.

[4]   Dong Sheqin, Bai Hongjie, Hong Xianlong, Bai Hongjie, and S. Goto, "Buffer planning for 3D ICs", *IEEE International Symposium on Circuits and Systems (ISCAS).*, 2009pp. 1735-1738 Taipei, Taiwan

[5]   S. Nakatake, K. Fujiyoshi, H. Murata, and Y. Kajitani, "Module packing based on the BSG-structure and IC layout applications", *IEEE Trans. Comput. Aided Des. Integrated Circ. Syst.*, vol. 17, no. 6, pp. 519-530, 1998.
[http://dx.doi.org/10.1109/43.703832]

[6]   H. Matsuda, S. Nakatake, and Y. Kajitani, "Optimum slicing-structure floor planning with routing area included", *IEICE Tech. Rep.*, vol. 94, no. 531, pp. 9-14, 1995.

[7]   T-C. Wang, and D.E. Wong, "An optimal algorithm for floorplan area optimization", *Proceedings of the 27th ACM/IEEE Design Automation Conference.*, 1990pp. 180-186
[http://dx.doi.org/10.1145/123186.123253]

[8]   D.F. Wong, and C.L. Liu, "A new algorithm for floorplan design", *23rd ACM/IEEE Design Automation Conference.*, 1986pp. 101-107 Las Vegas, NV, USA

CHAPTER 13

Underwater Bio-Mimic Robotic Fish

Ravi R.[1,*], **R. Tino Merlin**[1], **V. Harini Priya**[1], **T. Jerlin**[1], **U. Maheshwari**[1], **R. Indhu Rani**[1], **V. Brindha**[1] and **A. Celciya Effrin**[1]

[1] *Department of Electronics and Communication Engineering, Francis Xavier Engineering College, Tirunelveli, India*

Abstract: This chapter discusses the design and fabrication of biomimetic underwater robotic fish. A robot fish is a type of bionic robot that looks and moves like a real fish. Two motors, an Arduino microcontroller, Bluetooth, and a pump are required to complete the underwater robotic fish project. Motors are employed for quick forward and rotating motion, and the pump assembly aids in deep-water diving. In addition, sensors assist the robot in making intelligent judgments such as obstacle detection, direction shift, and so forth. Additionally, essential information such as live streaming, pressure, and temperature is provided. The innovative technology compromises the agility and performance of the robot that helps to achieve the real motion of the fish, making the robot competent for an aquatic-based design of the robot that helps to reduce the complex structure without applications such as underwater exploration, oceanic supervision, pollution level detection, and military detection This project is also beneficial.

Keywords: Arduino microcontroller, Bluetooth, Camera, Power supply, Servo motor, Temperature sensor, Water pump.

INTRODUCTION

A flexible tail mechanism allows a fish-like underwater robot to swim quicker and more softly while consuming less energy. The design of the robotic fish is based on the locomotion mechanism utilised in a variety of applications such as ocean development, military operations, and marine environment protection, and it necessitates a high-performance automated underwater vehicle (AUV). Robotics is the study of creating devices that can take the place of humans and mimic their actions. Robots can be employed in a variety of scenarios and for a variety of objectives, but many are now used in hazardous areas (such as radioactive material inspection, bomb detection, and deactivation), manufacturing operations, or in situations where humans are unable to live for example, in space, under-

---

* **Corresponding author Ravi R.:** Department of Electronics and Communication Engineering, Francis Xavier Engineering College, Tirunelveli, India; E-mail: directorresearch@francisxavier.ac.in

S. Kannadhasan, R. Nagarajan, Alagar Karthick, K.K. Saravanan & Kaushik Pal (Eds.)

water, in high heat, and clean up and containment of hazardous materials and radiation. Robots can take on any shape, but some are designed to seem like people. This is said to aid robot adoption in certain replicative activities that are normally performed by humans. Walking, lifting, speaking, cognition, and any other human activities are all attempted by these robots [1-5]. Many of today's robots are influenced by nature, making bio-inspired robotics a growing area. Today, we require a variety of technologies for monitoring and studying our environment. Underwater vehicles of this type have been developed in recent decades to investigate and experiment with water. Fish, an aquatic animal with outstanding man-ability and improved propulsion efficiency, is the best solution for underwater research. The two motors on this Robot's propulsion mechanism allow it to swim in the water. Sensors, actuators, and microprocessors, as well as a control mechanism, are all essential hardware and software components of underwater robots [6-10]. With its propelling mechanism, this robot is meant to swim in the water using two servo motors. It may also descend into the water thanks to a ballast tank system. We built and constructed an autonomous robotic fish to detect and follow an underwater object in this research as shown in Fig. (1).

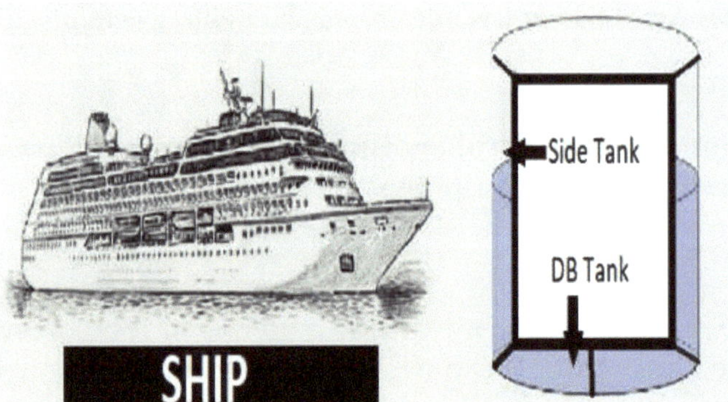

Fig. (1). Ballast Tank System.

Design of Underwater Robotic Fish

A streamlined head, a body, and a tail make up the most basic biomimetic robotic fish. All control modules, including a wireless communication module, batteries, and a signal processor, are housed in the head, which is often composed of a stiff plastic substance (fibreglass). The body could be made up of several jointed segments connected by servomotors. The rotation angle of the joint is controlled by servomotors. Pectoral fins are placed on both sides of the body in some designs to ensure stability in the water.

Motive power is provided *via* an oscillating caudal (tail) fin coupled to joints and operated by a motor as shown in Fig. **(2)**.

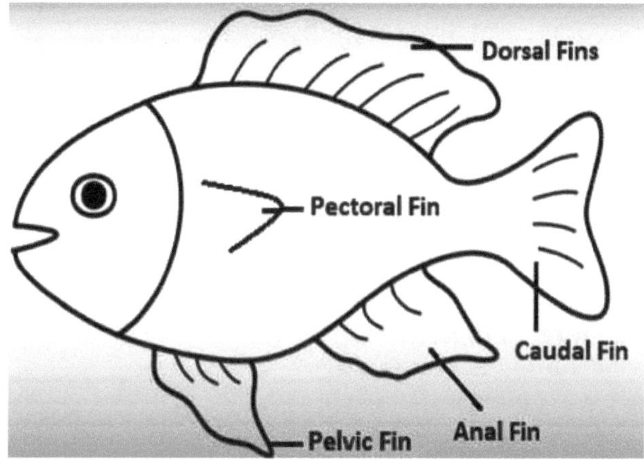

Fig. (2). Skeleton of robotic fish.

PRINCIPLE OF UNDERWATER ROBOTIC FISH

The most essential aspects of robot fish research and development are improving their control and navigation, allowing them to 'communicate' with their environment, travel along a certain course, and respond to directions to make their 'fins' flap. The three primary principles of underwater robotic fish are:

- Anguilliform: Propulsion by a muscle wave in the animal's body that goes from head to tail, similar to an eel.
- Carangiform: Salmon, Trout, Tuna, and Swordfish have oscillating tail fins and tail peduncles.
- Ostraciiform: Like the boxfish, it oscillates only the tail fin without moving the rest of the body.
- Mechanical design for underwater robot fishes can be divided into four types.
- Changing wave: The body wave is used to propel from head to tail. With so many hinges and joints, a smooth motion of the entire body is essential.
- Body Foil: The fish pushes the water away from them by oscillating its tail fin and moving its body in a wave pattern.
- Oscillating wing: This approach relies on an oscillating wing-shaped tail fin to provide nearly all propulsion force.
- Oscillating plate: In this manner, fish only vibrate at the tail fin, similar to a plate.

Fish Body Design

The purpose of this fish design is to efficiently biomimic the shape of streamlined fishes like tuna, which are swift aquatic swimmers in the oceans. The Tuna fish uses the BCF style of propulsion, which is one of two general forms of fish swimming: Body-CaudalFin (BCF) propulsion and median pectoral fin propulsion. The BCF mode is classified into four categories: Thunniform, Anguilliform, Subcarangiform, and Carangiform modes (Videler, 1993). The fraction of the body that actively engages in undulation differs between the modes. The carangiform way of swimming has been observed in tuna fish (BCF propulsion). Active undulation is restricted to the posterior one-third of the body in Carangiform phase. The body design has three major components: the anterior portion, the posterior portion, and the tail.

Mechanics of Underwater Robotic Fish

An underwater robot is a waterproof robot that can move in water in response to commands from the operator. There are two halves in this robot: one that transmits and one that receives. Acoustic signals (for example, sonar), light signals, electromagnetic signals, and bionic sensors are used in most underwater sensing systems. The position of a submerged object is determined by monitoring the transit duration and phase difference of acoustic pulses, which may operate over a much greater range and are unaffected by water turbidity. Sensing is done using bionic sensors. Platforms can bend, stretch, or morph into different shapes to sense environmental stimuli such as force, displacement, pressure, temperature, or chemicals by tiny mechanical deformations. Underwater electromagnetic signals can be used for navigation, sensing, and communications, among other things. A cable can provide short-range navigation while also reducing the required range for mobile communications. As a controller, an Arduino Nano board is used. The Mega 328 is a microcontroller with 32Kb of flash memory. The 12V LiPo battery provides power to the robotic fish.

Construction

The anterior half of the link has a profile made up of two circular arcs that mix with the flat area of the beginning of the middle part of the link, as shown in the illustration. This also eliminates any non-differentiability points at the point of contact between this component and the prior link's posterior part. Furthermore, the profiles have been chosen to avoid any places of non-differentiability on the link's surface. The smooth profile allows the wire to travel freely inside the grooves on the surface without causing too much friction. Furthermore, the wire moves perfectly along the link's surface, avoiding any spots where the wire loses contact with the link's surface, which would have thrown off the precise calculation of the wire lengths are substantially more complicated to run). However, in this scenario, the wire's running length equals the length of the grove's path, assuming perfect contact. Middle Element is the flat area of the connection that is characterised by this part of the link's design, which is pretty straightforward. Posterior Part is the link's posterior section and has two circular arcs that perform the same purpose as the link's anterior section. The wire glides across a grooved circular arc on the end section of the wire. The length of the wire is the only portion of the connection that is not fixed and changes depending on the angle between the two links.

Controller Design

Surge, sway, heave, pitch, and yaw are the five degrees of freedom available to the fish. The frequency of undulation of the tail fin, mean position of the servo-link mechanism, and movable mass displacement in the barycentre mechanism are used to manage them. The following paragraphs go over these control inputs in great depth. To begin, observe that this is an under-actuated system, as there are only three control inputs available to control a dynamical system with flexibility in five of the six dimensions conceivable. Assume that the robotic system is required to achieve a reference speed known as cruising speed. The real speed, on the other hand, begins at zero (rest) and attempts to stabilize at the cruising speed number by the action of the controller.

Sensors in Underwater Robotic Fish

For the propulsion mechanism with a specific angle of rotation, servo motors are employed. The servo motor angles for the fish's motility are controlled by PWM (Pulse Width Modulation) pulses. The water collection is powered by a DC motor. The adjustable servo motor angles regulate the fish's speed. The temperature is measured with an LM35. The built-in ADC converts analogue signals into digital data, which may then be sent to the base station through a wireless connection. The robotic fish has a camera attached to it. This camera aids in the monitoring and surveillance of the underwater environment.

- Photo Sensor: The fish will turn in the direction where the sensor detects the maximum fall of the light. If both sensors detect the same quantity of light, this indicates that the source of irradiance is straight ahead, and the robot will proceed.
- Pressure Sensor: The pressure sensor will be utilised to tell the robot how deep it is submerged, as well as to detect any agitation of the water in its immediate vicinity.

- Humidity Sensor: This device can be used to determine whether or not the robot is submerged in water.
- Audio Sensor: A microphone can be fitted to allow audio signals to be detected. A bigger surface area as well as many more microphones would be necessary to detect the position of the sound source as shown in Fig. (3).

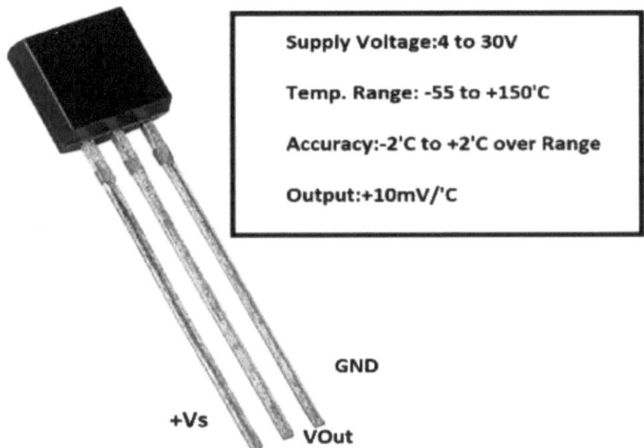

Fig. (3). LED Sensor.

Assumptions

1. Fish are active when there are no waves, this is, when the water is quiet and deep.

2. The fish is travelling at a constant positive velocity of U.

3. The coefficients of hydrodynamics are considered constant.

4. Off-diagonal terms in the relevant tensors mentioned ahead are ignored, implying that a decoupled dynamics model is used.

5. The control forces and moments outlined in the following section account for the influence of body flexibility.

6. Design factors are expected to passively stabilise roll.

Software Tracking

We created a vision-based tracking in our control software that processes the frames taken by the upper camera of the experimental setup in order to retrieve the position of the agents during an experiment and to operate the fish. To extract the moving items, each frame is subjected to a backdrop subtraction. For closed-loop control, the robotic fish poses are used to estimate the poses of the robotic fish. This program also controls the robotic fish caudal peduncle's beating. In terms of global architecture, robotic fish mobility is transmitted from the main application and contains the locomotion parameters. Each robotic fish has the characteristics mentioned in the following section implemented on-board. The robotic fish can emit events in the event of an impediment or power outage thanks to the event-based protocol. The IR proximity sensors are used to identify obstacles: when an IR signal received by the sensors exceeds a particular threshold, the robotic fish avoids the obstruction by rotating in the opposite direction of the received signal for a certain length of time. It enables the robotic fish to avoid colliding with objects such as walls and other robotic fish as shown in Fig. (**4**).

Fig. (4). Movements of Robotic Fish in Underwater.

Movement of Underwater Robotic Fish

Fish employs the following tools to swim efficiently:

- Fins are used to glide and turn.
- The air bladder allows you to move up and down.
- Body with a streamlined shape to reduce pressure drag.

Slime coat: To reduce dragging caused by friction.

Communication

Bluetooth is used to send and receive data from the robotic fish from the base station. Robotic fish may be operated manually or automatically using an Android application and a Bluetooth communication module. The data is transferred to the computer *via* the floating antenna for live streaming.

Proposed System

PVC is used to create the fish. PVC is a lightweight and durable material. Four servo motors power the robotic fish system. The movable joints are coupled to the servo motors. Two motors for the fish's right and left fins, as well as one for the body and another for the tail. For robotic fish locomotion, the body is made up of several jointed segments, and high torque servo motors are the major issue for position change and fast speed. It also provides the flexibility and stability required for stable mobility.

To detect objects, two infrared sensors are affixed to the fish's eyes, and a temperature sensor is attached to the fish's body. The robotic fish's drive circuit is designed on a PCB. Two lithium batteries are used to power all of these gadgets. The relay is also connected to the Arduino as well as the water pump.

Uses

Fish robots have recently become popular in a variety of applications, including maritime studies, military operations, and environmental protection. It necessitates high-performance autonomous underwater vehicles, particularly for propulsion, as well as significant advantages in terms of flexible maneuverability. The sensors on the mouth of the robotic fish can be used to study the oxygen levels in the water. It may learn about the various species in its surroundings by swimming among them and reporting on fish health.

- We can monitor water-bodies using this robotic fish.
- A video and picture capturing facility will aid in the observation of aquatic wildlife.
- Collecting water samples for water pollution monitoring is simple.
- This robotic fish can be equipped with a GPS tracker for navigation.
- This fish can also be equipped with an LED light for cleaning purposes.

CONCLUSION

Robotic fish have been created for this project to perform real-world tasks such as underwater object identification and tracking, navigation, and entertainment. A caudal fin, controlled and operated by a sensing circuit, servomotor, and computer algorithms, aids the robotic fish's mobility. Both mechanical and controller ideas are given in this study. We can make the posterior body flexible and light weight by installing servos at the head rather than at each joint. It has also been demonstrated how a flexor-extensor mechanism can be used to replicate the undulation of fish. Although the flexor-extensor mechanism has been utilised before in conjunction with the human arm to represent undulations and a variety of other motions, we were the first to employ it while lowering the number of servos. Robotic fish provide enormous potential for innovating new points of views. Around 70% of the world's aquatic bodies have yet to be explored. Exploring the waters is becoming increasingly vital as the world's population grows and puts pressure on terrestrial resources. Robotic fishes will undoubtedly be the stars of the show in regard to efficiently navigating the oceans and seas. The robot's two servo motors, which provide propulsion, enable it to actively manage degrees of freedom while also helping it to simulate the motion of a real fish. The robotic fish dives in smoothly and effectively thanks to the depth control mechanism created utilising a ballast tank.

REFERENCES

[1]   M. Sfakiotakis, D.M. Lane, and J.B.C. Davies, "Review of fish swimming modes for aquatic locomotion", *IEEE J. Oceanic Eng.*, vol. 24, no. 2, pp. 237-252, 1999.
[http://dx.doi.org/10.1109/48.757275]

[2]   E. Kanso, J.E. Marsden, C.W. Rowley, and J.B. Melli-Huber, "Locomotion of articulated bodies in a perfect uid", *J. Nonlinear Sci.*, vol. 15, no. 4, pp. 255-289, 2005.
[http://dx.doi.org/10.1007/s00332-004-0650-9]

[3]   M.J. Lighthill, "Note on Slender Fish Swimming", *J. Fluid Mech.*, no. 9, pp. 305-317, 1960.
[http://dx.doi.org/10.1017/S0022112060001110]

[4]   M.S. Triantafyllou, and G.S. Triantafyllou, "An efficient swimmingmachine", *Sci. Am.*, vol. 272, no. 3, pp. 64-70, 1995.
[http://dx.doi.org/10.1038/scientificamerican0395-64]

[5]   V. George, E.J. Anderson, E.J. Anderson, and J. Tangorra, ""Fish biorobotics: Kinematics and hydrodynamics of self-propulsion,"", *J. Exp. Biol.*, vol. 210, no. Pt 16, pp. 2767-2780, .

[6]   J.S. Martn, "An initial and boundary value issue modelling of fish-like swimming", *188.3 Rational Mechanics and Analysis Archive,* 2008.

[7]   "The creation of a self-contained robotic fish", *Biomedical Robotics and Biomechatronics,* pp. 1032-1037, 2012.

[8]   Z. Chen, S. Shatara, and X. Tan, "Modeling of a biomimetic robotic fish with a caudal fin made of an ionic polymer-metal composite", *IEEE/ASME Trans. Mechatron.*, vol. 15, no. 3, pp. 448-459, 2010.
[http://dx.doi.org/10.1109/TMECH.2009.2027812]

[9]   C. Rossi, J. Colorado, W. Coral, and A. Barrientos, "A revolutionary robotic fish design using SMAs to bend continuous structures", *Bioinspiration & Biomimetics,* 2011.

[10]   C.H. Le, Q.S. Nguyen, and H.C. Park, "A SMA-based actuation system for a fish robot", *Smart Struct. Syst.*, vol. 10, no. 6, pp. 501-515, 2012.
[http://dx.doi.org/10.12989/sss.2012.10.6.501]

CHAPTER 14

IoT-Based Automatic Irrigation System

Raja M.[1,*], **N.M. Nithish**[1], **Saravana Shankar B.**[1] and **Sadhurwanth D.**[1]

[1] Department of Electronics and Instrumentation Engineering, Kongu Engineering College, Perundurai, Erode, Tamil Nadu, India

Abstract: Agriculture is one of the backbones of our Indian economy. India is primarily an agricultural country. It plays an important role in the development of our nation. This project proposes an automatic irrigation system, because it maintains the moisture content present in the soil by automatic irrigation system. This setup uses a capacitive soil moisture sensor v1.2 that measures the exact amount of soil moisture. It monitors soil properties such as temperature, humidity, soil moisture, and motor status. These parameters are measured using a soil moisture sensor, a DHT11 sensor, which is controlled by a NodeMCU that acts both as a microprocessor and as a server. It is possible to remotely control many farm operations from any part of the world through IoT.

Keywords: DHT 11 sensor, IoT, NodeMCU, Pump, Relay, Soil moisture sensor.

INTRODUCTION

The automatic irrigation Developing Model is a real-time monitoring system. Most farmers use many hectares of farmland. It's becoming very difficult to monitor and check every area of large land. Sometimes there is a chance of uneven irrigation for crops. This results in low quality crops, which also leads to a loss of production. At present, automation is among the most essential roles in human life. This not only brings comfort, but it also reduces energy consumption or saves time. Industries are now using control and automation machine that is expensive and not approximate for use in farming [1, 2]. This also constructs a low-cost, intelligent irrigation technology that can be used by Indian farmers. This project proposes an automated irrigation system and maintains the moisture

---

* **Corresponding author Raja M.:** Department of Electronics and Instrumentation Engineering, Kongu Engineering College, Perundurai, Erode, Tamil Nadu, India; E-mail: mraja4@gmail.com

S. Kannadhasan, R. Nagarajan, Alagar Karthick, K.K. Saravanan & Kaushik Pal (Eds.)

content of the soil through automatic irrigation. In the existing farming system, crops are monitored with the help of Arduino boards and GSM technology [3], in which Arduino boards act as microcontrollers but not as servers. Therefore, in order to overcome all of these features, a microcontroller like NodeMCU is the latest version that acts as both a microcontroller and a server. The main feature of this methodology is that its cost of installation is cheap and has numerous advantages. Here you can access and visualize and control the farming system on your laptop, cell phone, *etc.* [4, 5].

LITERATURE SURVEY

Before briefing IoT-based Automated Irrigation System, we shall revise the existing agricultural system. In a few of the existing irrigation systems, soil moisture is monitored using a voltage meter and water irrigation controller system .In few other irrigation systems, fuzzy logic controllers are used for the effective irrigation of various crops. Fuzzy logic system rises the standard rate of the values obtained and helps to make decisions. Wireless Sensor Network and a GPRS Module are used for automatic irrigation in the current system of irrigation [6-8]. The temperature and humidity are measured using sensors and controlled by a microcontroller.

The WIU also has a GPRS module that transmits data to the server on a public mobile internet. The information data achieved can be monitored remotely through a graphical representation of the server. This irrigation system makes it easy to monitor and cultivate even in places with less water, and making the system feasible. But Zigbee Protocol makes this system costly. Modern agriculture is based on a greenhouse which must be accurately controlled both in terms of humidity and temperature. The atmospheric conditions around the plants vary from place to place, making it nearly impossible to irrigate uniformly in in all locations of the field manually. To that end, GSM technology is used to gain the report of the irrigation process *via* farmers, mobile phone, or other social communication devices [9-11].

PROPOSED SYSTEM

The main objective of this study is to supply water when there is less moisture content in the soil without anyone's presence to avoid excess water supply during irrigation and also monitoring soil parameters such as soil moisture, atmospheric temperature and atmospheric humidity. The various parameters of the field can also be remotely controlled at any time, through mobile phone and web applications. This sends signals to the mobile phone whether to supply water or not (*i.e.,* when the field is dry) to the field. It includes Node MCU and sensors such as soil moisture sensor and DHT11 sensor, relays, OLED display and submersible motor.

METHOLOGY

The current methodologies include Arduino board and GSM technology. It operates only as a microcontroller, but never as a server. Thus, to overcome these differences, a microcontroller like NodeMCU which is the latest version that acts as both a microcontroller and a server is used. Fig. (**1**) shows the block diagram that gives the basic outcome of the project. It also includes components used in the project such as soil moisture sensor, humidity sensor, and temperature sensor in a single sensor known as DHT 11 sensor, OLED display, relay, and submersible pump for use in this experiment.

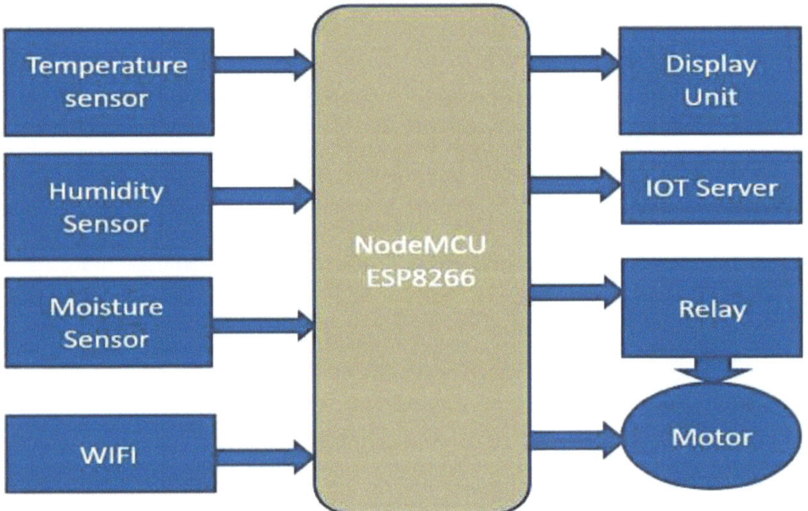

Fig. (1). Block diagram.

HARDWARE DISCRIPTION

NodeMCU

NodeMCU works on a multiparadigm, lightweight, high-level programming language designed primarily for embedded applications known as Lua and specifically used for IoT-based applications. It works both as a microcontroller and as a modem. It also works with the help of ESP8266 [NodeMCU] Wi-Fi and hardware equipment depending on the ESP-12 module.

RELAY

Relay is the most frequently used switching device in electronics. Relay has an input terminal set and an operating terminal set. The traditional shape of the relay uses an electromagnet to close and open the contacts. Coil 1 is used to trigger the relay. Normally, one end is in contact with 5 V and the other end with the ground.

Coil 2 is used to trigger the relay, usually, one end is connected to 5 V and the other end to the ground. The relay is used to control the submersible pump that is connected to the controller and the sensor.

DHT 11 SENSOR

The DHT 11 sensor is used to measure both the temperature and humidity of the atmosphere. It consists of two parts, the capacitive humidity sensor and a thermistor, which helps to measure both atmospheric humidity as well as atmospheric temperature. The digital temperature and humidity sensor is of low cost. It is very easy to use our project. And it has one main advantage it measures the values of these parameters precisely every 3 to 5 seconds. Atmospheric humidity alters the capacitance of the DHT11 sensor and gives us an accurate humidity output. There is also a small chip in the sensor that helps convert the analog signal to the digital output. The DHT11 sensor is connected to the NodeMCU and the display for the reference of the produced reading. It has three pins for connecting to the circuit, *i.e.,* VCC, data pin, and ground pin.

Analog Capacitive Soil Moisture Sensor

Capacitive soil moisture sensor which helps to measure the soil moisture with respect to the change in capacitance. That is, the moisture of the soil changes the capacitance of the rod and produces a certain amount of voltage with respect to the change in capacitance. It basically produces about 1.2V to 3.0V. Capacitive soil moisture sensor works based on the principle of change in capacitance. It measures the change in capacitance rather than the change resistance.

Submersible Pump

The submersible pump is used to supply water to the soil under the control of the microcontroller. The relay helps control the pump when the pump is switched on and off. The submersible pump works with an analog capacitive soil moisture sensor. Once the soil moisture level is below a certain level, the motor is automatically switched on. Water is pumped out of the reservoir and supplied to the soil. When the soil moisture level reaches a certain limit, the pump switches off automatically with the support of the microcontroller and the soil moisture sensor is connected to the pump.

The submersible pump operates on a supply of 3.3V to 6V. This can pump approximately 120 liters per hour even when the supply is low (*i.e.,* 220 mA). The pump is submerged in water and the tube is connected to the pump outlet and is powered to make the pump work in any condition.

OLED Display

The display is blue-colored 0.96-inch OLED (Organic Light-Emitting Diode) display unit. With the help of SPI/IIC protocol, we can interface the display module with any microcontroller. It has a resolution of 128×64. The package consists of three main objects, a display, a display board, and a pre-soldered 4-pin male header on board. It is a self-light-emitting technology collected from a multilayered thin organic film which is placed between a cathode and anode system. Unlike Liquid Crystal Display technology, the OLED display does not require any backlight. It has a high application potential for practically every type of display and is considered the latest technology for the next generation of flat panel displays. We use the OLED display to display the parameters that we are about to measure. This works under the control of the connected devices as well as the controller unit.

IoT Cloud

The IoT cloud used in this project is a ThingSpeak server, and it is an IoT(Internet of things) based cloud platform that allows to receive and analyse sensor data from the cloud and develop IoT applications. The Internet of Things is an emerging trend technology where more number of built-in devices (things) connected to the Internet. These digital technologies share information with other users and other places but often facilitate sensor information to cloud storage as well as cloud computing resources where information is processed and evaluated for important insights. Affordable cloud computing has expanded internet connectivity to make this trend possible. IoT solutions are designed for a wide range of various applications, including health surveillance, vehicle fleet tracking, industrial monitoring and control, environmental monitoring and management, and remote monitoring.

Setting up of Thinkspeak Server

The following procedures should be followed to set up the Thingspeak server: create an account with https://thingspeak.com/ and login and fill in the details in the account.

By clicking the "Channel" button, we can create a new channel by filling these following details. Click the API Key to see the "Write API Key" button. Insert the API key. We can click the "Private View" button and modify the window display as you desire; this is all a part of the Thingspeak IoT cloud.

Now coming to the program section of the IoT-based Automatic Irrigation System Source Code of NodeMCU ESP8266, you can upload the code directly to the development board. Before all this, remember to add OLED Display Library Adafruit GFX Library, and SSD1306 Library. By changing the API Key of the Thinkspeak, Wi-Fi SSID and password, the program is executed.

INTERFACING & WORKING

A schematic diagram of the IoT-based Automatic Irrigation System project is shown. This is where the Fritzing tool is used to design a circuit, where the soil moisture device is connected to the A0 pin of NodeMCU and the DHT11 pin to the D4 pin. The relay is controlled by a D5 pin of NodeMCU. An OLED Display is connected to a NodeMCU 12c pi. The relay and motor are controlled by 5v over the VIN pin of NodeMCU. The DHT11 Sensor, the capacitive Soil moisture sensor V1.2, and the OLED display are controlled by a 3.3v supply. Capacitive soil moisture sensor, DHT11 sensor is interfaced with NodeMCU where the air temperature, air humidity, soil moisture and motor status can be monitored through thinkspeak.com.

The OLED display will start displaying real-time data as soon as the device is powered. When the soil moisture content present in the soil is reduced, the water pump will turn on and irrigate the soil until the required moisture level is reached. All of this can be remotely monitored from anywhere in the world through the use of a Thinkspeak server [2].In this image, it shows that the water level of the soil is being nurtured and the limit of the soil moisture has been set. So the water pump automatically turns off and the supply to the motor is shut. The observations along with the soil moisture level as well as temperature, humidity, and their respective graphs and motor status are also shown on the IoT platform. In Fig. (**2**) the IoT platform shows us the exact condition of the parameters and the status of the sensor without delay in the readings and the graph. Farmers can visualize and analyze online data from any other part of the world [1].

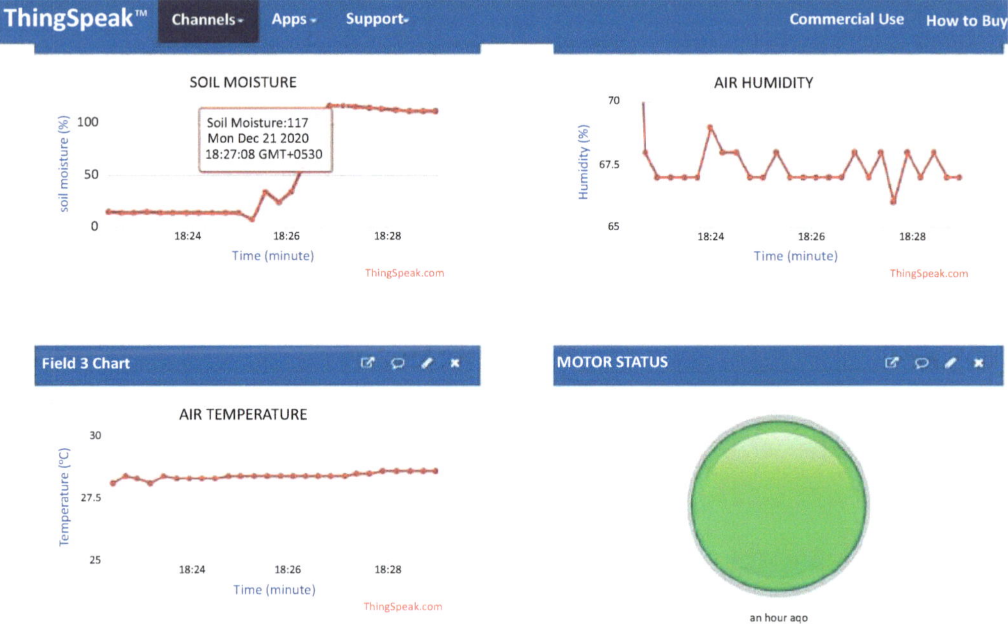

Fig. (2). Readings & graph viewed through IoT.

RESULTS & CONCLUSION

Automatic irrigation of water is carried out successfully in all parts of the field whenever a low amount of soil moisture is sensed, the engine switches automatically ON and the soil becomes wet, the engine turns off, the prototype developed in this project prevents human work and water wastage. By implementing this smart set-up, crops will increase productivity. This is more compatible and economical in regard to setting up on a large scale. A real-time data is shown. Whenever the soil moisture levels are reduced, a water pump switches ON and irrigates the area till the required moisture has been reached.

In Fig. (**3**), the final output of our project is water irrigation that pumps the water automatically and provides water to the soil in water depletion while there is no manual supply of water. Fig. (**4**). shows the automatic switching off pump with the help of a microcontroller and relay and maintaining the soil moisture.

Fig. (3). shows the final output of the project.

Fig. (4). Shows the automatic control of water supply.

Data can be accessed online from anywhere through a Thingspeak server . We can visualize the private view of the server. Here we can monitor these 4 parameters, soil moisture, temperature, and humidity conditions as well as relay status. Farmers can analyse and visualize the generation of parameters, and they can also be monitored.

REFERENCES

[1]   C Amardeo, and I G Sarma, "Identity in the Future Internet of Things", *Wireless Pers Commun,* vol. 49, pp. 353-363, 2009.

[2]   S.R. Das, S. Chita, N. Peterson, B.A. Shirazi, and M. Bhadkamkar, "Home automation and security for mobile devices", *IEEE PERCOM Workshop,* pp. 141-146, 2011.
[http://dx.doi.org/10.1109/PERCOMW.2011.5766856]

[3]   I. Gautam, and S.R.N. Reddy, "Innovative GSM based Remote Controlled Embedded System for Irrigation", *International Journal of Computer Applications,* vol. 47, no. 13, 2012.

[4]   Y. Kim, R.G. Evans, and W.M. Iversen, "Remote sensing and control of an irrigation system using a distributed wireless sensor network", *IEEE Transactions on Instrumentation and Measurement,* vol. 57, no. 7, pp. 1379-1387, 2008.

[5]   Y.G. Lin, "An Intelligent Monitoring System for Agriculture Based on ZigBee Wireless Sensors Network Journal", *Advanced Materials Research, Manufacturing Science and Technology,* vol. 383-390, pp. 4358-4364, 2008.

[6]   M.T. Batte, ""Changing computer use in agriculture: Evidence from Ohio"", *Computers and Electronics in Agriculture.,* vol. Vol. 47, no. 1, pp. 1-13, 2005.

[7]   S.D.T. Kelly, N.K. Suryadevara, and S.C. Mukhopadh, "Towards the Implementation of IoT (Internet of things) for Monitoring Environmental Conditions in Homes", *IEEE Sensors Journal,* vol. Vol. 13, no. 10, pp. 3846-3853, 2013.

[8]   B.R. Shiraz Pasha, "Microcontroller Based Automated Irrigation System", *The International Journal Of Engineering And Science (IJES),* vol. 3, no. 7, pp. 06-09, 2014.

[9]   S.R. Kumbhar, and P. Arjun, "Microcontroller based Controlled Irrigation System for Plantation", *Proceedings of the International MultiConference of Engineers and Computer Scientists,* vol. Volume II, 2013

[10]   "IoT based Automated Irrigation System", *IJSRD - International Journal for Scientific Research & Development,* vol. 3, no. 4, 2015.

[11]   V. Naga, and R. Gunturi, "Micro Controller Based Automatic Plant Irrigation System", *International Journal of Advancements in Research & Technology,* vol. 2, 2013.

SUBJECT INDEX